国家出版基金项目
NATIONAL PUBLICATION FOUNDATION

中国最美古树

The Most Beautiful Ancient Trees in China

《国土绿化》杂志 主编

By *Land Greening* Magazine

中国画报出版社·北京

China Pictorial Press · Beijing

图书在版编目（CIP）数据

中国最美古树：汉英对照 /《国土绿化》杂志主编
. -- 北京：中国画报出版社，2021.5
 ISBN 978-7-5146-1990-4

Ⅰ.①中… Ⅱ.①国… Ⅲ.①树木 - 中国 - 图集
Ⅳ.① S717.2-64

中国版本图书馆 CIP 数据核字 (2020) 第 271859 号

中国最美古树
《国土绿化》杂志 主编

出 版 人：于九涛
项目统筹：齐丽华
责任编辑：刘晓雪
特约审稿：张跃平
英文翻译：王国振
英文编辑：朱露茜　叶淑君　陈星宇　郭　菲
装帧设计：罗家洋
责任印制：焦　洋
出版发行：中国画报出版社
地　　址：中国北京市海淀区车公庄西路 33 号　邮编：100048
发 行 部：010-68469781　010-68414683（传真）
总编室兼传真：010-88417359　版权部：010-88417359
开　　本：16 开（787mm x 1092mm）
印　　张：11.5
字　　数：200 千字
版　　次：2021 年 5 月第 1 版　2021 年 5 月第 1 次印刷
印　　刷：北京汇瑞嘉合文化发展有限公司
书　　号：ISBN 978-7-5146-1990-4
定　　价：98.00 元

大地守望者——中国古树名木

古树名木是中华民族悠久历史与文化的象征，具有十分重要的历史、文化、科学、经济和生态价值。每一棵古树名木都具有丰富的历史文化内涵，都蕴含着丰富的人文情怀。

每一个民族的文化复兴，都是从总结自己的遗产开始的。古树名木是我国民族文化的重要象征，是不可再生的文化遗产，是人类社会历史发展的佐证，具有重要的科学、历史、人文与景观的价值。在中华民族五千年历史的长河中，我们的祖先不仅创造了辉煌灿烂的文化，还培植了数量众多、饱经沧桑的古树名木，历代名人重视植树护树，与古树有着不解之缘，见树如见人，从夏商周、秦汉唐宋、元明清至今，都留下了各朝各代的名人植树。这些古树历尽沧桑，数经风雨，如今枝繁叶茂，备受世人敬仰，它们是在5000多年文明发展中孕育的中华优秀传统文化的代表，对延续和发展中华文明、促进人类文明进步，发挥着重要的作用。

我国古树资源丰富、种类众多。据不完全统计，目前全国古树数量超过1000万棵。书中收录了85株全国各地的珍贵名木。这些树全部出自中国林学会评选出的"中国最美古树"。为了保证数据的准确性，每株古树的树龄及树的胸（地）围等相关数据均来自2016年"最美古树"推选活动中各地相关部门提供的树木档案。

这些中国最美古树中，既有见证上古传说的轩辕黄帝手植柏，也有被誉为"植物活化石"的"中国一号水杉母树"；既有西北大漠中守望千年的"神树"——胡杨，也有南国水乡里独木成林的"鸟的天堂"；既有树围巨大、需近十人才能合抱的西藏巨柏，也有树形轻盈、姿态婀娜的江南流苏树。中国最美古树千姿百态，各美其美，共同构成了美丽中国的宏大叙事。

Ancient Precious Trees in China

Ancient precious trees, symbolic of the long history and culture of the Chinese nation, have very important historical, cultural, scientific, economic and ecological values. Each possesses profound historical and cultural connotations and rich humanistic feelings.

Every nation has started its cultural renaissance beginning by summing up its own heritage. As the important symbol of national culture, ancient famous trees represent a cultural heritage, as well as the evidence of the historical development of human society, which, however, is nonrenewable. These trees boast important scientific, historical, humanistic and landscape values. In the 5,000 years of history of the Chinese nation, our ancestors created a splendid culture in many forms. During the period, they cultivated a large number of old and famous trees that have experienced many vicissitudes. Celebrities of various ages attached high importance to planting and protecting trees, and developed close ties with them. These ancient trees remind people of these celebrities in history. Today, we still have trees left behind from the dynasties of Xia (2070–1600 BCE), Shang (1600–1046 BCE) and Zhou (1046–256 BCE), Qin (221–207 BCE), Han (202 BCE–220 CE), Tang (618–907), Song (960–1279), Yuan (1271–1368), Ming (1368–1644) and Qing (1644–1911) to today. Despite the passage of time, these trees remain robust, standing proudly in face of all weathers. With luxuriant branches, they enjoy high respect of people today as they are representatives not only of a culture stretching back five millennia, but play an important role in the continuation and development of Chinese civilization and the advancement of human civilization.

Incomplete statistics show there are more than 10 million ancient trees in China. The book contains 85 ancient precious and famous trees found all over the country. All these trees belong to the "Most Beautiful Ancient Trees in China" selected by Chinese Society of Forest. To ensure the accuracy of the data, relevant information on the age and circumference of each tree are all from files given by relevant local departments in the "Most Beautiful Trees" selection activity in 2016.

Of all the beautiful trees, there is the cypress planted by the Yellow Emperor Xuanyuan (2697–2599 BCE), which witnessed the rise of this ancient legendary ruler; the No.1 Mother Tree of Metasequoia Glyptostroboides in China, known as a "living fossil of plants"; the diversiform-leaved poplar, a sacred tree that has been watching the passage of thousands of years in the northwest desert; the Bird's Paradise in the water villages of southern China; Giant Cypress in Tibet, which is so thick that ten people linking hands are required to encircle it; and enchanting Chinese fringe trees unique to the area south of the Yangtze River. They join others to show a beautiful China.

目 录 Contents

002 　最美黄山松：迎客松
　　　The Most Beautiful Mt Huangshan Pine: The Pine Greeting Guests

004 　最美油松：九龙松
　　　The Most Beautiful Chinese Pine: Nine-dragon Pine

007 　最美马尾松：鹿角松
　　　The Most Beautiful Chinese Pine: Staghorn Pine

008 　最美水松："天下第一水松"
　　　The Most Beautiful Chinese Cypress "The Unparalleled"

010 　最美金钱松："全国第一条儿"
　　　The Most Beautiful Pseudolarix Kaempferi: "The No. 1 Long Strip in China"

012 　最美红松：虎松
　　　The Most Beautiful Korean Pine: Tiger Pine

015 　最美长白松：美人松
　　　The Most Beautiful Changbai Pine: Beauty Pine

016 　最美罗汉松：东晋古罗汉松
　　　The Most Beautiful Arhat Pine: The Ancient Arhat Pin in the Eastern Jin Dynasty

018 　最美白皮松：九龙松
　　　The Most Beautiful Lacebark Pine: Nine-Dragon Pine

020 　最美小叶杨：九龙蟠杨
　　　The Most Beautiful Small-leaf Poplar: Nine-Dragon Poplar

022 　最美胡杨：额济纳胡杨林中的"神树"
　　　The Most Beautiful Diversiform-Leaved Poplar: Divine Tree of the Diversiform-Leaved Poplar Forest in Ejina

024 　最美枫杨：神农架枫杨
　　　The Most Beautiful Chinese Ash: A Chinese Ash in Shennongjia Forest

026 　最美白榆："兄弟榆"
　　　The Most Beautiful White-Bark Elm: "Brother Elms"

028 　最美大果榆："夫妻树"
　　　The Most Beautiful Ulmus Macrocarpa Hance: "Husband-Wife Tree"

031 　最美杉木："伞树"
　　　The Most Beautiful China Fir: "Umbrella Tree"

032	最美柳杉："柳杉王"	
	The Most Beautiful Cryptomeria Fortunei: "King of Cryptomeria Fortunei"	
034	最美南方红豆杉：松阳红豆杉	
	The Most Beautiful Chinese Yew: The Chinese Yew in the Songyang County	
036	最美东北红豆杉：东北红豆杉"树王"	
	The Most Beautiful Taxus Cuspidata: "Tree King"	
039	最美长苞铁杉：长苞铁杉王	
	The Most Beautiful Tsuga Longibracteata: Tsuga Longibracteata King	
040	最美油杉："谊父树"	
	The Most Beautiful Chinese Fir: "Adoptive Father Tree"	
042	最美铁坚油杉："夫妻树"	
	The Most Beautiful *Keteleeria Davidiana*: "Husband and Wife Trees"	
044	最美水杉：利川"水杉王"	
	The Most Beautiful Chinese Redwood: "Chinese Redwood King" in Lichuan	
047	最美侧柏：黄帝手植柏	
	The Most Beautiful Chinese Arborvitae: Cypress Planted by the Yellow Emperor	
049	最美圆柏：曹魏古柏	
	The Most Beautiful Sabina: Beautiful Ancient Sabina of the Wei State of the Three-Kingdoms Period	
050	最美大果圆柏：热振森林公园大果圆柏	
	The Most Beautiful China Savin: Big-Fruit Tree in Razheng Forest Park	
052	最美国槐："天下第一槐"	
	The Most Beautiful Scholar Tree: "the Best in the World"	
055	最美樟树：德化"樟树王"	
	The Most Beautiful Camphor Tree: "King of Camphor Trees" in Dehua	
056	最美榕树："小鸟天堂"	
	The Most Beautiful Banyan Tree: "Birds Paradise"	
059	最美绿黄葛树："上甲贵古榕树群"	
	The Most Beautiful Ficus Virens: "Ancient Banyan Woodlot in Shangjia of Guizhou"	
060	最美观光木：大丘"观光木树王"	
	The Most Beautiful *Tsoongiodendron Odorum*: "King of *Tsoongiodendron Odorum*" in Daqiu	

063　最美南紫薇：莲花山南紫薇
The Most Beautiful *Lagerstroemia Indica*: *Lagerstroemia indica* in Lotus Hill

064　最美紫薇：印江紫薇
The Most Beautiful Crape Myrtle: Yinjiang Crape Myrtle

066　最美闽楠：永安"百年神树"
The Most Beautiful Phoebe Bournei: Yong'an's "100-Year-Old Divine Tree"

068　最美桢楠："桢楠王"
The Most Beautiful Phoebe Zhennan: "King of Phoebe Zhennan"

070　最美玉兰：玉兰花谷玉兰树
The Most Beautiful Yulan Magnolia: Yulan Magnolia in Magnolia Flower Valley

072　最美重阳木：芷江"云树"
The Most Beautiful *Bischofia Polycarpa*: "Cloud Tree" in Zhijiang

075　最美桂花："九龙桂"
The Most Beautiful Sweet-Scented Osmanthus: "Nine-Dragon Osmanthus"

077　最美檫木：沙县"仙姑树"
The Most Beautiful Sassafras Tzumu: the "Fairy Tree" in Shaxian County

078　最美香榧："中国香榧王"
The Most Beautiful Chinese Torreya: "King of Chinese Torreya"

080　最美百日青：临海百日青
The Most Beautiful Bamboo Leaf Pine Overlooking the Sea

082　最美蚬木：龙州蚬木王
The Most Beautiful *Burretiodendron Hsienmu*: a Tree King in Longzhou

084　最美湖北梣："映泉鸳鸯树"
The Most Beautiful *Fraxinus hupehensis*: "Yingquan Sweethearts Tree"

086　最美新疆野苹果："树龄最长的野生苹果树"
The Most Beautiful Wild Apple Tree in Xinjiang: "The Oldest Wild Apple Tree"

089　最美胡桃："核桃树王"
The Most Beautiful Walnut Tree: "King of Walnut Trees"

090　最美三球悬铃木："其娜尔"古树
The Most Beautiful *Platanus Orientalis* Linn.: "Chinar" Ancient Tree

092　最美梓叶槭：金家冲梓叶槭
The Most Beautiful *Acer Catalpifolium* Rehd: *Acer Catalpifolium* Rehd in Jinjiachong

094　最美黄连木：陇南黄连木
The Most Beautiful *Pistacia Chinensis* Bunge: *Pistacia Chinensis* Bunge in Longnan

097　最美亮叶水青冈："千手观音"
The Most Beautiful *Fagus Lucida*: "Thousand-hand Kwanyin"

098　最美流苏树：连云港"糯米花树"
The Most Beautiful Chinese Fringe Tree: "Glutinous Rice Popcorn Tree" in Lianyungang

101　最美荔枝：漳州"桂枝"
The Most Beautiful *Litchi* Tree: "Sweet-scented Osmanthus *Litchi* Tree" in Zhangzhou

102　最美木棉：中山堂木棉
The Most Beautiful Kapok Tree: Kapok Tree in the Sun Yat-sen Memorial Hall

104　最美古梅："潮塘宫粉"
The Most Beautiful Ancient Plum: "Powder in Chaotang"

106　最美枳椇：南雄拐枣
The Most Beautiful Raisin Tree: Nanxiong Honey Raisin Tree

109　最美米椎："岭南第一大椎"
The Most Beautiful Mizhui Tree: "The Largest Mizhui Tree in Lingnan"

110　最美人面子："人面子王"
The Most Beautiful *Dracontomelon Duperreanum* Pierre: "King of *Dracontomelon Duperreanum* Pierre Trees"

112　最美板栗：怀柔古板栗
The Most Beautiful Chinese Chestnut Tree: Ancient Chestnut Tree in Huairou

115　最美酸枣：北京"酸枣王"
The Most Beautiful Wild Jujube: "King of Wild Jujube" in Beijing

116　最美槲树："菜树奶奶"
The Most Beautiful *Quercus Dentata*: "Grandma of Oak Trees"

118　最美青檀："檀公古树"
The Most Beautiful Wingceltis: "Ancient Wingceltis Tree"

120　最美七叶树：安康七叶树
The Most Beautiful Chinese horse-chestnut (*Aesculus chinensis* Bunge) in Ankang

122 最美沙梨："梨树王"
The Most Beautiful Sand Pear Tree: "King of Pear Trees"

124 最美文冠果：渭南文冠果
The Most Beautiful Shiny-leaved Yellowhorn Tree: Shiny-leaved Yellowhorn Tree in Weinan

127 最美辽东栎："虎龙古树"
The Most Beautiful Liaole Oak: "Tiger-Dragon Ancient Tree"

128 最美大果榉：邯郸大果榉
The Most Beautiful *Zelkova Sinica*: *Zelkova Sinica* in Handan

130 最美黄栌：泽州"分脉树"
The Most Beautiful Smoke Tree: 'Divided Tree' in Zezhou

132 最美巨柏：西藏柏树王
The Most Beautiful Cypress: King of Cypress in Tibet

134 最美沙棘：西藏古沙棘树
The Most Beautiful Sea Buckthorn: Ancient Sea Buckthorn Tree in Tibet

136 最美桑树：西藏"古桑树王"
The Most Beautiful Mulberry Tree: King of Ancient Mulberry Trees in Tibet

138 最美楸树：原平"龙凤古楸"
The Most Beautiful Catalpa Bungei: Ancient Twin Catalpa Trees in Yuanping

140 最美麻栎：永济橡树
The Most Beautiful Sawtooth Oak: Oak Trees in Yongji

142 最美元宝槭：和顺元宝槭
The Most Beautiful Acer Truncatum: Acer Truncatu in Heshun County

145 最美紫丁香："华北最大紫丁香"
The Most Beautiful Lilac: Largest Lilac in North China

146 最美银杏：莒县银杏
The Most Beautiful Ginkgo: Juxian Ginkgo

148 最美红花天料木：霸王岭母生
The Most Beautiful Homalium Hainanense: Homalium Hainanense in Bawangling National Forest Park

150 最美陆均松："五指神树"
The Most Beautiful Dacrydium Pierrei: " Five-Finger Sacred Tree"

152 　最美冬青：南阳冬青树
　　　The Most Beautiful Holly Tree: The Holly Tree in Nanyang

154 　最美皂荚：商丘皂角树
　　　The Most Beautiful Honey Locust: Locust Tree in Shangqiu

156 　最美蒙古栎：千年蒙古栎
　　　The Most Beautiful Quercus Mongolica: Millennium Mongolian Oak

158 　最美青杨：清原青杨
　　　The Most Beautiful Cathay Poplar: Cathay Poplar in Qingyuan

160 　最美赤松：江东赤松王
　　　The Most Beautiful Red Pine: King of Red Pine in Jiangdong

162 　最美剑阁柏：剑阁翠云廊古柏
　　　The Most Beautiful Jiange Cypress: Ancient Cypress in Jiange Cuiyun Corridor Scenic Area

164 　最美香果树："丁木大仙"
　　　The Most Beautiful Emmenopterys Henryi: 'Dingmu Grand Immortal'

166 　最美柏木：七贤古柏
　　　The Most Beautiful Cedarwood: Seven-Sage Cypress

168 　最美雅安红豆树："红豆仙树"
　　　The Most Beautiful Red Bean Tree: "Red Bean Fair Tree"

170 　最美高山榕："华夏榕树王"
　　　The Most Beautiful Banyan: "King of Chinese Banyan"

最美黄山松：迎客松

迎客松树寿逾千年，为黄山特级保护古树名木，先后入选黄山十大名松之首、安徽省名木、"中国最美黄山松"等；也是世界之名贵木种，列入世界文化与自然遗产名录，是中国唯一上榜"全球最著名16棵树木"的古树。

迎客松挺立于海拔1670米的玉屏楼青狮石旁。其上部枝叶虬结平密，树冠如幡似盖，且偏向一侧，下部有两大侧枝横空斜出，既展现欲揽五湖四海、迎送八方宾朋的雍容俊美的姿态，又似颔首展臂向游人致意，天然神态，令人叫绝。

迎客松之名始见于清咸丰九年（1859年）歙县人黄肇敏《黄山纪游》。根据现有资料，迎客松首次被摄入镜头的时间是1912年10月，为著名画家汪采白所摄。1929年，太平县人陈少峰主编《黄山指南》选用迎客松照片刊于全书正文之前，迎客松从此广为世人所知。1959年，巨幅铁画《迎客松》被悬挂在北京人民大会堂安徽厅里，党和国家领导人多次在铁画前与外国客人合影留念，使其成为中国人民同世界人民友好的象征，蜚声中外。1994年，人民大会堂东大厅也悬挂了黄山籍画家刘晖所作国画《迎客松》，它见证了我国与世界各国人民所结下的深厚友谊。

自1981年起，黄山管理部门为迎客松设立了一个独一无二的岗位——"守松人"，这是迄今为止全中国乃至全世界唯一的专职树木"警卫"。

The Most Beautiful Mt Huangshan Pine: The Pine Greeting Guests

The Pine Greeting Guests, with a tree lifespan of 1,000 years, is listed as an old and famous tree under special protection on Mt Huangshan. It is regarded as surmounting the Top 10 Mt Huangshan Famous Pines, the famous pine in Anhui Province, and the "Most Beautiful Mt Huangshan Pine." As a precious tree, it is listed in the World Cultural and Natural Heritage List. It is the only ancient tree of China on the list of the "16 of the Most Magnificent Trees in the World."

The Pine Greeting Guests stands by the stone lion of the Jade Screen Pavilion at an altitude of 1,670 meters. With dense upper branches and leaves, the Pine Greeting Guests has a crown leaning to one side, under which are two lateral branches on the lower part; this impresses people who see the pine as bowing and stretching out its arms to greet friends and guests. The natural expression is amazing.

The name of "The Pine Greeting Guests" was first found in the *Mt Huangshan Travelogue* written by Huang Zhaomin, a native of Shexian County of Anhui Province in 1859, or the ninth year of the reign of Qing Dynasty Emperor Xianfeng. According to the available data, The Pine Greeting Guests was first photographed by famous painter Wang Caibai in October of 1912, founding year of the Republic of China (1912–1949). In 1929, Chen Shaofeng, a native of Taiping County, compiled a book entitled *Mt Huangshan Guide*, Wang Caibai's photo of *The Pine Greeting Guests* was selected for the title page. From then on, making The Pine Greeting Guests known to all in China. In 1959 when the Great Hall of the People was built in Beijing, a giant portrait of *The Pine Greeting Guests* was hung in its Anhui Hall. When Party and government leaders meet and have a picture taken with visiting guests, they have posed in front of this portrait many times — a symbol of friendship between the Chinese people and the people of the other parts of the world. This also helps spread the fame of The Pine Greeting Guests. In 1994, the East Hall of the Great Hall of the People also hung the traditional Chinese painting *The Pine Greeting Guests* by Liu Hui, a Huangshan painter, witnessing the deep friendship between China and the peoples of the rest of the world. *The Pine Greeting Guests* is held as a national treasure.

Since 1981, the Huangshan Management Department has set up a unique post guarding for The Pine Greeting Guests, which is the only full-time "tree guard" in China and even in the world.

中文名：黄山松

拉丁名：*Pinus taiwanensis* Hayata

所在地：安徽省黄山市黄山风景区玉屏楼景区

树龄：1000 年

胸（地）围：219 厘米

树高：1020 厘米

冠幅（平均）：1200 厘米

Chinese Name: 黄山松

Latin Name: *Pinus taiwanensis* Hayata

Location: Huangshan Jade Screen Pavilion Scenic Spot, Huangshan City, Anhui Province

Tree Age: 1,000 years

Chest (Floor) Circumference: 219 cm

Tree Height: 1,020 cm

Crown Width (Average): 1,200 cm

最美油松：九龙松

此树位于河北省承德市丰宁满族自治县五道营乡三道营村。它奇特美观，具有丰富的历史文化内涵，珍贵珍稀程度和保护价值都很高。从外观看，主干之上九条枝干横斜逸出，皆水平生长，每枝弯曲向前，枝头好像龙头，枝身弯弯曲曲犹如龙身，树皮呈块状，好似龙鳞，九条枝干条条像龙，飞腾而起，故当地百姓称其为"九龙松"。

九龙松号称"天下第一奇松"，它有四奇：从远处看，古松好像一个大的盆景，变换角度景象则迥然各异。从北面朝南看，树的轮廓和长势，同对面的驸马山山脉的轮廓走向一模一样。尤为奇特的是，古树层次分明，错落有致地分为上下4层，与其相对应的驸马山也相应地分为上下4层，山与树遥相呼应，自成情趣，此为第一奇。九龙松枝干长势奇特，经过500年的岁月，此树并没有向上长，而是向四处延伸生长，其枝干全部盘旋、弯曲再翻转着生长，此为第二奇。古松尽管每年结出松果，但松子在未成熟时洁白饱满，成熟后里面却只剩一层皮，此为第三奇。第四奇是古松独木成林，微风掠过，松涛阵阵。

The Most Beautiful Chinese Pine: Nine-dragon Pine

This tree is located in Sandaoying Village of Wudaoying Township, Fengning Manchu Autonomous County in Chengde City of Hebei Province. This unique and beautiful tree has rich historical and cultural connotations, with a high degree of rarity and protection value. Nine branches on the main trunk grow horizontally. Each branch twists and turns with the tip looking like a dragon's head. Pieces of bark look like dragon scales. Altogether, the pine has nine branches, hence its name.

The Nine-dragon Pine is known as "the most wonderful pine in the world." It has four peculiar features: First, when looked at from a distance, the ancient pine looks like a large bonsai, which changes its appearance when viewed from different angles. When viewed from the north, the pine looks exactly the same as the opposite-facing Fuma Mountain in terms of the outline and the growing trend. What this means is that the ancient pine has four layers in response to the shape of Fuma Mountain. Second, the 500-year-old pine does not grow straight upward; instead, its branches grow in all directions but spiral, bend and turn over. Third, it produces pinecones every year. The pine nuts are white and plump when not yet ripe; and there is only one layer of skin left after maturity. And, the fourth peculiarity is that it whistles like an entire forest even in the gentlest breeze.

中文名：油松
拉丁名：*Pinus tabuliformis* Carr.
所在地：河北省承德市丰宁满族自治县五道营乡三道营村
树龄：500 年
胸（地）围：330 厘米
树高：600 厘米
冠幅（平均）：2600 厘米

Chinese Name: 油松
Latin Name: *Pinus tabuliformis* Carr.
Location: Sandaoying Village of Wudaoying Township, Fengning Manchu Autonomous County in Chengde City, Hebei Province
Tree Age: 500 years
Chest (Floor) Circumference: 330 cm
Tree Height: 600 cm
Crown Width (Average): 2,600 cm

最美马尾松：鹿角松

该树位于福建省屏南县岭下乡葛畲村苏氏祖坟的正上方，主干1.5米高处一分为三，三根并排笔直的树干直径都在1米以上，高耸入云。从远处看，三根树干的形状酷似一根根"鹿茸"，挺拔翘立，因此，古树被当地人形象地称为"鹿角松"。

据传，葛畲苏氏始祖奶泰、奶顺兄弟二人当年从建安忠溪随母迁入时，寄人篱下，家境贫困，历尽艰辛。苍天不负有心人，奶泰、奶顺兄弟艰苦谋生，克勤克俭，家道日渐殷实，但也常感父亲、叔叔等尸骨尚无合适地方掩埋，心中不免唏嘘。于是兄弟二人邀请风水先生在村旁找了一穴墓地，地形好似一只活泼可爱的麒鹿。俗有"麒鹿出洋"之称，两人大喜，遂将父亲、叔叔等前辈尸骨合葬于此，了却一桩心愿。为美化墓地，苏氏先祖在坟墓周围播撒了松子，奇怪的是，只在墓正中上方长出了一株。这株松树经过葛畲苏氏历代宗亲的精心呵护，终长成今天的参天奇树。

（图片摄影　庄晨辉）

The Most Beautiful Chinese Pine: Staghorn Pine

The tree is located right above Su's ancestral tomb in Geshe Village, Lingxia Township in Pingnan County of Fujian Province. The tree trunk rises 1.5 meters and then separates into three parallel straight trunks each being one meter in diameter and towering into the clouds. When viewed from a distance, the three trunks look like three "velvet antlers." This is why the pine is named "Staghorn Pine" by locals.

Legend has it that two brothers—Nutai and Nushun, the first ancestors of the Su family in Geshe—moved here with their mother from Zhongxi, Jian'an. With the wolf at the door, they underwent many hardships. Thanks to the hard work, diligence and thrift of the two brothers, however, the family gradually gained wealth. The two brothers invited a Fengshui geomancer to find a graveyard for their deceased father and uncle. The graveyard that was finally found looked like a lovely unicorn deer was held as a good omen for the family. Their father and uncle were then buried there. Su's ancestors planted spread many pine seeds in the area where the graveyard was located, but, strangely, only one grew in the center of the tomb. The pine tree, after the careful care of successive generations of Geshe Su family, finally grew into a skyscraping tree as we see today.

(Photos by Zhuang Chenhui)

中文名：马尾松
拉丁名：*Pinus massoniana* Lamb.
所在地：福建省屏南县岭下乡葛畲村
树龄：1206 年
胸（地）围：502.4 厘米
树高：2580 厘米
冠幅（平均）：1580 厘米

Chinese Name: 马尾松
Latin Name: *Pinus massoniana* Lamb.
Location: Geshe Village, Lingxia Township, Pingnan County, Fujian Province
Tree Age: 1,206 years
Chest (Floor) Circumference: 502.4 cm
Tree Height: 2,580 cm
Crown Width (Average): 1,580 cm

最美水松："天下第一水松"

该树位于福建省漳平市永福镇李庄官洋自然村小溪畔，有"天下第一水松"的美誉。据传，这棵古水松是五代十国时期，李庄（当时称饶寨）先民饶氏种的风水树。今饶氏虽已在李庄消失，但宋末北方客裔到此定居后，历代后人亦对此树珍爱有加，规定在树的附近不准搭盖建筑物。旧时，中元节乡人还自发到此整治环境，保护生态。

古水松根系近百米（有农夫在劳作时于远处挖到此树大根），虽屡遭雷击伤痕累累，但仍顽强屹立。每年春至，新叶长出，犹如一位披着青纱的慈祥老人，庇护着子孙后代。古水松主干心腐成洞，很早以前，敬放着树神宝像，让乡人供奉。

距这株老水松 2.5 千米处的后孟村小溪边，还有一株胸径 1.4 米，树高 23.8 米，平均冠幅 10 米，传说树龄 600 年的"子水松"，与李庄水松遥相呼应。相传，古时八仙之一的铁拐李到李庄省亲时，发现老水松奇异，就想多给它留一枝根脉，特地折一水松枝，插在不远相对处，吹口仙气，让它茁壮成长，遂成今日之景观。

（图片摄影 庄晨辉）

The Most Beautiful Chinese Cypress: "The Unparalleled"

The tree is located by the brook of Lizhuang Guanyang Village, Yongfu Town in Zhangping City of Fujian Province. Legend has it that this ancient Chinese cypress was a Fengshui tree planted by the Rao family, ancestors of Lizhuang (then called Rao stockaded village) during the period of the Five Dynasties (907–960) and Ten Kingdoms (902–979). The Rao family has long since disappeared in Lizhuang. When businesspeople from the north settled here at the end of the Song Dynasty (960–1279), later generations cherished the tree by stipulating that no buildings were allowed to be built nearby. In the old days, during the Zhongyuan Festival on the 15th day of the seventh lunar month, villagers spontaneously came to clean up the environment and protect it.

The root system of the ancient Chinese cypress extends nearly 100 meters (according to a farmer's discovery of a big root of this tree at that distance while working). Although the tree has been repeatedly damaged by lightning, it still stands firm. Every spring, new leaves sprout, just like a kind old man in green yarn, protecting future generations. A long time ago, a treasured statue of the tree god was placed in the hole of the trunk for worship.

Some 2.5 km away from the old Chinese cypress is also a tree with DBH of 1.4 meters, tree height of 23.8 meters, and average crown width of 10 meters. It is said that when Tieguai Li, one of the Eight Immortals of Taoism who limps about with an iron walking stick, came to Lizhuang for a visit, he was amazed to find the old Chinese cypress and took one of its branches and planted it nearby. He breathed on it and the branch gradually grew into a junior Chinese cypress.

(Photos by Zhuang Chenhui)

中文名：水松
拉丁名：*Glyptostrobus pensilis*（Staunt.）Koch
所在地：福建省漳平市永福镇李庄官洋自然村
树龄：2100 年
胸（地）围：697.1 厘米
树高：2600 厘米
冠幅（平均）：1670 厘米

Chinese Name: 水松
Latin Name: *Glyptostrobus pensilis*（Staunt.）Koch
Location: Lizhuang Guanyang Village, Yongfu Town, Zhangping City, Fujian Province
Tree Age: 2,100 years
Chest (Floor) Circumference: 697.1 cm
Tree Height: 2,600 cm
Crown Width (Average): 1,670 cm

最美金钱松："全国第一条儿"

此树生长于浙江省杭州市临安区，浙江天目山国家级自然保护区"大树王国"景区的中心地带，位于原狮子正宗禅寺（现称开山老殿）前面悬崖下沟谷中。

古树生长旺盛，主干通直匀称，神似影视剧《西游记》中孙悟空手持的金箍棒。二十世纪六七十年代，天目山林场场长见其长于深沟，树干笔直直冲蓝天，便为它起名"冲天树"。由于树干太高，如果戴着帽子仰望树顶的话，稍不小心帽子就会掉到地上，又有人形象地称之为"脱帽树"。杭州人把瘦瘦长长的高个子称为"条儿"，以树喻人，金钱松可算是天目山这个"森林王国"里的"条儿"。据了解，目前国内还没有第二株金钱松能有如此高度，因此便有了"全国第一条儿"的美名。

金钱松又名金松，在分类上属松科金钱松属植物，在全世界仅有一种，列入中国植物红皮书和国家二级重点保护植物名录，为世界著名的五大庭园观赏树种之一。

中文名：金钱松

拉丁名：*Pseudolarix amabilis*（Nelson）Rehd.

所在地：浙江省杭州市临安区天目山自然保护区

树龄：660 年

胸（地）围：322 厘米

树高：5800 厘米

冠幅（平均）：1500 厘米

Chinese Name: 金钱松

Latin Name: *Pseudolarix amabilis*（Nelson）Rehd.

Location: Tianmu Mountain Nature Reserve, Lin'an District, Hangzhou, Zhejiang Province

Tree Age: 660 years

Chest (Floor) Circumference: 322 cm

Tree Height: 5,800 cm

Crown Width (Average): 1,500 cm

The Most Beautiful Pseudolarix Kaempferi: "The No. 1 Long Strip in China"

This tree grows in Lin'an District of Hangzhou City, Zhejiang Province, in the center of the "Big Tree Kingdom," a scenic spot of the Zhejiang Tianmu Mountain National Nature Reserve. It is located in the valley under the cliff in front of the Kaishan Old Hall, originally the Lion Buddhist Temple.

The ancient tree grows vigorously and its trunk is straight and symmetrical, looking like the golden cudgel used by the Monkey King in TV play *Journey to the West*. In the 1960s and 1970s, the general manager of the Tianmu Mountain Forest Farm named it "Shooting up to the Sky Tree" because the tree, located in a deep valley, rises into the sky. The trunk is so high that one's hat would fall off if trying to look up at its crown from ground level. So, the tree is also called "Hat Falls off Tree." Hangzhou people call one who is tall and thin "long strip." So, the tree in the forest kingdom is also called Long Strip. It is understood that at present, there is no second Pseudolarix Kaempferi in China that can rival the "long strip."

Pseudolarix Kaempfer is also called an Umbrella Pine, belonging to the category of Pseudolarix Kaempferi of Pinaceae, being the only one in the world. It has been listed in the *China Plant Red Data Book* and the list of national second-class key protected plants. It is one of the five famous garden ornamental trees in the world.

最美红松：虎松

在黑龙江省海林市长汀镇这个举世闻名的中国雪乡太平沟林场67林班原始林景区里，生长着一棵十分罕见的大红松。此树因胸围粗壮、树干笔直、观赏性高而尤为珍稀。据测算，这棵红松树龄已有600余年，被当地誉为"红松树王"，又被称为"虎松"。

说起这棵树王，还有一段神奇的传说。清光绪年间，一群闯关东的山东人来到这深山老林中采集人参，为首的名为孙达。他刚来时雄心勃勃，一定要采到上等人参（棒槌），不料，他领着几个人在山上转悠了半个月竟一无所获，吃的也没有了。他和几个伙计正为采不到人参而郁闷时，忽然狂风大作，电闪雷鸣，大雨滂沱，只见一只小虎崽儿在一棵树下瑟瑟发抖，好像在向他求救。他毫不犹豫地冲了出去，把它抱在怀里回到地窨子。这时雷电更大了，把地窨子都震塌了，孙达也被震晕了，但他始终把小虎崽儿抱在怀里没有放松。雨过天晴后，孙达苏醒了，但他怀里的小虎崽儿却不见了，袖口里有一个白绢，上面写着四行字"棒槌要采撷，需拜山神爷，难事要弄明，山顶大红松。"他和大家一说，有人说老虎就是山大王。大家不敢怠慢，立即奔向山顶，果然有一棵硕大的红松树屹立在山顶。孙达选择一个吉日良辰，把上好的祭品供上，隆重祭拜山神。说来灵验，祭拜后每次放山，都收获颇丰。他们感谢山神、感谢树王，便把这棵大红松尊称为"虎松"，并在附近做了一个小牌坊，经常去拜谒祈求顺利、平安。

The Most Beautiful Korean Pine: Tiger Pine

There is a very rare Korean pine tree in the primitive forest of the Number 67 Forest Squad of the Taipinggou Forest Farm in Changting Town of Hailin City in Heilongjiang Province. This tree is very rare because of its thick chest, straight trunk and high ornamental value. According to the calculation, it is more than 600 years old, known locally as "the king of Korean pines," also known as a "tiger pine."

Legend has it that, during the reign of Qing Emperor Guangxu (1875–1908), a group of Shandong people had made their way here to collect ginseng in this deep mountain. They were led by a man named Sun Da, who vowed to collect the first-class ginseng (mallet). When they were depressed by the lack of ginseng after half-a-month work, they suddenly saw a tiger cub shivering under a tree as if asking for help. He rushed to take it into a nearby cellar. The thunder and lightning became stronger, and the cellar collapsed. Sun Da was knocked unconsciousness, but he still held the tiger cub in his arms. When he woke up, he found the tiger cub in his arms had left, leaving behind a white piece of silk with words reading: "To pick the mallet, you need to worship the God of the Mountain with a huge Korean pine atop." He and his men lost no time to rush to the top of the mountain and did find a giant Korean pine at its top. Sun Da chose an auspicious day and offered the best sacrifice to the Mountain God. Believe it or not, each worship activity on the Mountain God resulted in bumper harvest. To thank the Mountain God, they named the Korean pine "Tiger Pine" and erected a small memorial archway for worship.

中文名：红松
拉丁名：*Pinus koraiensis* Sieb. et Zucc.
所在地：黑龙江省牡丹江市海林市长汀镇太平沟林场
树龄：600 年
胸（地）围：432 厘米
树高：3754 厘米
冠幅（平均）：1050 厘米

Chinese Name: 红松
Latin Name: *Pinus koraiensis* Sieb. et Zucc.
Location: Taipinggou Forest Farm, Changting Town, Hailin City, Heilongjiang Province
Tree Age: 600 years
Chest (Floor) Circumference: 432 cm
Tree Height: 3,754 cm
Crown Width (Average): 1,050 cm

中国最美古树
The Most Beautiful Ancient Trees in China

中文名：长白松

拉丁名：*Pinus sylvestris* L.var. *sylvestriformis*

所在地：吉林省延边州安图县二道白河镇红石林场

树龄：350 年

胸（地）围：329 厘米

树高：2300 厘米

冠幅（平均）：1700 厘米

Chinese Name: 长白松

Latin Name: *Pinus sylvestris* L.var. *sylvestriformis*

Location: Hongshi Forest Farm, Erdaobaihe Town, Antu County, Yanbian Prefecture, Jilin Province.

Tree Age: 350 years

Chest (Floor) Circumference: 329 cm

Tree Height: 2,300 cm

Crown Width (Average): 1,700 cm

最美长白松：美人松

此树位于吉林省延边州安图县二道白河镇红石林场石峰景区内。长白山曾有过3次规模较大的火山爆发，分别是1597年（明万历二十五年）、1668年（清康熙七年）和1702年（清康熙四十一年），"美人松"经历了1668年和1702年两次火山大爆发，幸存至今。

古松主干高大，挺拔笔直，下部枝条早已脱落，侧生枝条全都集中在树干的顶部，形成绮丽、开阔、优美的树冠，而那些左右伸出的修长枝条既苍劲而又妩媚，微风吹拂之下，轻轻摇曳，仿佛在招手致意，因此当地居民又叫它"美人松"。

吉林省白河林业局为保护这一珍贵树种，专门成立了"美人松保护管理处"，并于2006年移交长白山保护开发区管委会，使得这一濒危物种得以被很好地保护。"美人松"现为国家一级重点保护野生植物（国务院1999年8月4日批准）。

在《长白山纪行》中有这样一段描述："妩媚的美人松羞花闭月，枝条酷似妙龄少女的香臂，舒展开去，潇洒脱俗。叶冠犹如美人的秀发，光彩照人，文雅迷人。"美人松的外表虽像闺中少女，可它的内里却潜藏着与一切恶劣环境抗争的顽强毅力。

The Most Beautiful Changbai Pine: Beauty Pine

This tree is located in the Shifeng scenic area of Hongshi Forest Farm, Erdaobaihe Town, Antu County in Yanbian Prefecture of Jilin Province. There were three large-scale volcanic eruptions in Changbai Mountain, respectively in 1597 (the 25th year of Emperor Wanli reign in the Ming Dynasty), 1668 (seventh year of Emperor Kangxi reign in the Qing Dynasty) and 1702 (the 41st year of Emperor Kangxi reign in the Qing Dynasty). The Beauty Pine experienced two large-scale volcanic eruptions in 1668 and 1702.

The trunk of the old pine is tall and straight, but the lower branches have already fallen off. The lateral branches are all concentrated at the top of the trunk, forming a beautiful crown. The slender branches stretch out from the left and right and sway gently in wind as if waving to viewers. The local residents call it "Beauty Pine."

In order to protect this precious tree species, the Baihe Forestry Bureau of Jilin Province established the Beauty Pine Protection and Management Office, which was transferred to the Management Committee of the Changbai Mountain Protection and Development Zone in 2006, making this endangered species well-protected. Beauty Pine is now under national key protection approved by the State Council on August 4, 1999.

It is recorded in *Travel Notes of Changbai Mountains*: "The Beauty Pine possesses incomparable beauty causing the flowers to blush and the moon to hide. Its branches are like the fragrant arms of a young beauty... its crown is like a beauty's hair, shining and elegant." Although the Beauty Pine is like a girl in the boudoir, it has a strong will to fight against all the travails of a harsh environment.

■ 最美罗汉松：东晋古罗汉松

"演径捣药已无踪，古观丹崖翠壁重。要识庐山先辈面，含情一抚六朝松。"这是清代诗人商盘游历庐山时留下的诗句，诗中"六朝松"即指东林寺内的罗汉松，传说为东晋高僧慧远大师手植，历经六朝，故称"六朝松"。无独有偶，在庐山五大丛林之一的万杉寺东北侧，也长有一棵植于东晋中期的古罗汉松。

古罗汉松枝繁叶茂，苍虬上耸，树影婆娑。据考证，这株古树属东晋中期遗物，树龄近 1600 年，被誉为"赣北罗汉王"。在佛教中自己得道的人被称为罗汉，这株千年的罗汉松亦有罗汉的样貌。树干被一条条虬龙盘绕着，从下往上升腾，好似几条龙在嬉戏玩耍，赫然就是一幅天然的九龙戏珠图，一节一节的突起仿佛龙的纹饰。古罗汉松枝叶茂密，远望如一把巨伞，雄浑苍劲，傲然挺拔。这株罗汉松不结果，也许它不必用繁衍后代来证明自己的存在。

罗汉松契合中国文化"长寿"的寓意，当地村民在树旁建了一个小小的庙，期盼这棵有灵的千年罗汉松荫佑村民。据村里人介绍，上海有一客商曾经三度想以 250 万元的高价买走此树，但村民并不为其所动，当地人爱树护树的精神由此可见一斑。

The Most Beautiful Arhat Pine: The Ancient Arhat Pin in the Eastern Jin Dynasty

The Qing Dynasty (1644–1911) poet Shang Pan once visited Lushan, a mountain in the northern part of Jianxi Province, and wrote a poem that began: "All literati who have visited the mountain marvel at the Six-Dynasty Pine." It is said it was planted by Master Huiyuan, an eminent monk in the Eastern Jin Dynasty (417–420). It continued to grow through six successive dynasties, thus its current name. Coincidentally, in the northeast of Wanshan Temple, one of the five major forest retreats in the Lushan area, there is also an ancient Arhat pine planted during the middle period of the Eastern Jin Dynasty.

The ancient Arhat pine has luxuriant branches and leaves. According to textual research, it is nearly 1,600 years old and is known as "The king of Arhats in northern Jiangxi Province." In Buddhism, one who attains the highest state of spiritual enlightenment in known as an "arhat." The pine has the appearance of an arhat with the tree trunk being encircled by "dragons" rising from the bottom, providing a natural picture of nine dragons playing a with a pearl. From a distance, it looks like a giant umbrella. Strangely, it does not bear pine seeds, showing it does not need to produce "offspring" as it has long and flourishing life.

Arhat pine is in line with the meaning of "longevity" in Chinese culture. Local villagers have built a small temple beside the tree in the hope it can protect them. According to the villagers, a businessman in Shanghai tried to buy the tree for 2.5 million Yuan on three occasions, but the villagers refused out of their infinite love for the tree.

中文名：罗汉松
拉丁名：*Podocarpus macrophyllus*（Thunb.）D.Don
所在地：江西省九江市星子县白鹿乡万杉村
树龄：1600 年
胸（地）围：760 厘米
树高：1700 厘米
冠幅（平均）：1000 厘米

Chinese Name: 罗汉松
Latin Name: *Podocarpus macrophyllus*（Thunb.）D.Don
Location: Wanshan Village, Bailu Township, Xingzi County, Jiujiang City, Jiangxi Province
Tree Age: 1,600 years
Chest (Floor) Circumference: 760 cm
Tree Height: 1,700 cm
Crown Width (Average): 1,000 cm

最美白皮松：九龙松

此树位于北京市门头沟区永定镇戒台寺风景区戒台殿门口，胸围6.5米，平均冠幅23米，树高18米。

白皮松又名虎皮松、白骨松、白果松等，为常绿乔木，是我国特有树种，树形多姿，苍翠挺拔，别具特色，枝轮生，冬芽显著，芽鳞多数，覆瓦状排列，是园林绿化和庭院绿化的优良树种，近年已闻名世界。此株古白皮松为戒台寺十大名松之首，冠幅巨大，雄伟壮观，因九条主干延伸，如九条银龙凌空飞舞，又似九条神龙在守护着戒坛，故得名"九龙松"。

该树为唐武德年间所植，至今已1300多年历史，是北京地区同树种最古老的一株，也是我国乃至世界"古白皮松之最"。

The Most Beautiful Laceback Pine: Nine-Dragon Pine

The tree is located at the gate of Jietai Hall of the Jietai Temple in Yongding Town, in Mentougou District of Beijing. It has a chest circumference of 6.5 meters, an average crown width of 23 meters and a height of 18 meters.

Laceback Pine is also known as Tiger Skin Pine, White Bone Pine and White Nut Pine, being an evergreen arbor tree species that is a unique species in China. It has a variety of tree shapes, and is tall and straight. Its branches appear in rotation, winter buds are obvious and numerous bud scales are arranged in a tile shape. It has been adopted for garden landscaping and courtyard greening projects, and has been well-known all over the world in recent years. This ancient laceback pine tops the list of the 10 famous pines in the Jietai Temple. As it has a huge crown and looks so magnificent with nine trunks extending like nine silver dragons flying in the air and nine divine dragons guarding the altar, it is also known as the "Nine-Dragon Pine."

The tree was planted in the reign of Emperor Wude of the Tang Dynasty more than 1,300 years ago. It is the oldest tree of the same species in the Beijing area, and is also the "most ancient white bark pine" in China and the world.

中文名：白皮松
拉丁名：*Pinus bungeana* Zucc.ex Endl.
所在地：北京市门头沟区永定镇戒台寺风景区
树龄：1300 年
胸（地）围：650 厘米
树高：1800 厘米
冠幅（平均）：2300 厘米

Chinese Name: 白皮松
Latin Name: *Pinus bungeana* Zucc.ex Endl.
Location: Jietai Temple Scenic Area, Yongding Town, Mentougou District, Beijing
Tree Age: 1,300 years
Chest (Floor) Circumference: 650 cm
Tree Height: 1,800 cm
Crown Width (Average): 2,300 cm

最美小叶杨：九龙蟠杨

在河北省承德市平泉县柳溪镇下桥头村，有一株小叶杨，当地人称"九龙蟠杨"，树冠覆盖面积750平方米，可谓"独木成林"。此树具有一个特殊的现象，从东南方向看，树冠轮廓酷似中国陆地版图，十分壮观，让人拍案叫绝。

这棵树一株三干，粗者好像老象之腿，稳如磐石，最细的也粗如铁柱，刚劲若骨。枝枝分杈犹如游龙，有的矫首昂视，睥睨众生；有的俯首低眉，张开双臂，似欢迎远客到访；有的像戏曲名家甩起的长长水袖；有的如蛟龙探海，九曲连环；还有的压肩叠背，互相扶持。此树独木成林，秋风掠过，树叶沙沙作响，像是为游客弹奏一曲意蕴深远的古琴曲。

关于这棵古树，当地文学资料中还记载着这样一个传说：辽代开泰九年（1020年），一年一度的"秋捺钵"（辽国皇帝秋猎于山）开始了，圣宗皇帝耶律隆绪与皇后萧氏到马盂山（今辽河源光秃山）打猎，见一只梅花鹿在一棵杨树上磨角，皇帝张弓搭箭欲射，皇后不忍杀戮这个小生灵，抢先拉弓虚发，小鹿闻风而逃，皇帝心领神会，与皇后相视而笑。此后，这棵杨树在皇后箭矢的痕迹处，中央主干一分为三，长成了9个枝杈，蜿蜒成9条形态各异的虬龙，"九龙蟠杨"因此得名。

为更好地保护古树，平泉对"九龙蟠杨"实行封闭管理，设立了围栏，鼓励当地群众参与古树保护工作。

The Most Beautiful Small-leaf Poplar: Nine-Dragon Poplar

In Xiaqiaotou Village, Liuxi Town in Pingquan County of Chengde City, Hebei Province, there is a small leaf poplar, known locally as the "Nine-Dragon Poplar." Its crown covers an area of 750 square meters, which can be described as "a single tree large enough to be a forest by itself." This tree has a special phenomenon. When viewed from the southeast, the crown contour of the tree is very similar to the land map of China, which is very spectacular and amazing.

The tree has three trunks. The thick one is like an old elephant's leg, stable as a rock, and the thinnest is as thick as an iron pillar. Its branches are like flying dragons, some looking up to heaven and down on all living beings; and are some bowing their heads and lowering their eyebrows and opening their arms to welcome visitors from afar. Some are like the long sleeves thrown up by famous dramatists, some like dragons exploring the sea with nine tunes, and others support each other. The tree itself is so huge that it is like a forest on its very own, where tree leaves give out melodious sounds like piano playing.

There is a legend about this ancient tree in local literature. In 1020, the annual autumn hunting expedition was underway during the period of the Liao State (907–1125). Liao King Shengzong Yelelungxu and Empress Xiao went hunting in Mayu Mountain (now Guangtu Mountain, water source of the Liaohe River). They suddenly spotted a sika deer grinding its horn on a poplar tree. The king brought out his bow and arrow, ready to shoot. The queen hated to see such a lovely animal killed and shot an arrow at a tree before her husband did. The sika was alarmed and fled from the smiling couple. The trunk of the tree hit by the queen's arrow later branched out into three sub-main trunks each with three branches. The nine branches twisted like nine dragons climbing around the tree. hence the Nine-Dragon Poplar.

In order to better protect such ancient trees, Pingquan exercises closed management over the Nine-Dragon Poplar by setting up a fence around it. And local people are encouragedto participate in the protection of all ancient trees.

中文名：小叶杨
拉丁名：*Populus simonii* Carr.
所在地：河北省承德市平泉县柳溪镇下桥头村
树龄：500 年
胸（地）围：1960 厘米
树高：2000 厘米
冠幅（平均）：2690 厘米

Chinese Name: 小叶杨
Latin Name: *Populus simonii* Carr.
Location: Xiaqiaotou Village, Liuxi Town, Pingquan County, Chengde City, Hebei Province
Tree Age: 500 years
Chest (Floor) Circumference: 1,960 cm
Tree Height: 2,000 cm
Crown Width (Average): 2,690 cm

最美胡杨：额济纳胡杨林中的"神树"

在渺无边际的大漠戈壁深处，有一片充满传奇色彩的神秘土地——内蒙古自治区额济纳旗，这里干旱多风、降水稀少。然而，就是在这样一片以沙漠戈壁为主的恶劣环境中，却有着世界三大胡杨林之一——额济纳胡杨林。

额济纳胡杨林中，生长着一棵被当地人称为"神树"的胡杨树。"神树"位于内蒙古自治区阿拉善盟额济纳旗达来呼布镇以北25千米处。

相传，300多年前，土尔扈特人来到额济纳草原，发现了这棵胡杨树。一日，王爷的夫人想做一只奶桶，就令工匠锯下树南侧的一根枝干。不料，奶桶做成了，夫人的左脚大拇指却溃烂了，任凭怎样用药，伤口就是不愈。后来，有一名高僧到这棵树下仔细观察，终于发现了胡杨神。于是高僧召集众僧侣，诵经七天七夜，夫人的脚伤才好。人们都说："这棵树真的有神啊。"于是，土尔扈特人怀着崇敬的心情将此树供奉为神树。时至今日，每当冬末春初、青黄不接的季节，远远近近的牧人们都会来到这棵神树前虔诚地诵经祷告，祈求风调雨顺草畜兴旺。

胡杨是当今世界上最为古老的杨树品种，被誉为"活着的化石树"。"神树"正是额济纳44万亩胡杨的典范。如今，在"神树"周围30多米的范围内，从它发达的根系中，又蘖生出5棵茁壮的胡杨，牧人们把它们叫做"母子树"。"神树"的家族茁壮茂盛，巍然耸立在沙丘红柳间。

The Most Beautiful Diversiform-Leaved Poplar: Divine Tree of the Diversiform-Leaved Poplar Forest in Ejina

In the boundless Gobi Desert is a mysterious land full of marvelous legends —namely, the Ejina Banner of the Inner Mongolia Autonomous Region. Dry and windy, the part of the world has little precipitation. In such a harsh environment dominated by the Gobi Desert, is one of the three largest diversiform-leaved poplar forests in the world.

In the diversiform-leaved poplar forest of Ejina is one called a "divine tree" by local people. It is located 25 km north of Dalai Hubu Town, Ejina Banner, Alashan League of Inner Mongolia Autonomous Region.

Legend has it that, more than 300 years ago, Torghut people came to Ejina Grassland where they found this diversiform-leaved poplar tree. One day, the wife of the Torghut king wanted to have a milk bucket made and so ordered the craftsman to fell a branch of the diversiform-leaved poplar on the southern side of the tree. The milk bucket was made, but the left thumb of the lady became ulcerated and, no matter how the medicine was used, the wound never healed. An eminent monk examined the tree carefully and finally found the Diversiform-Leaved Poplar God who was unhappy. The eminent monk summoned all the monks to chant sutras for seven days and seven nights until the lady's food was healed. From then on, Torghut people worshiped this tree as a sacred one. Today, at the end of winter and the beginning of spring, Torghut herdsmen come to worship the divine tree in pray for the prosperity of the grassland and their livestock in the coming year.

Diversiform-leaved poplar is the oldest poplar species in the world, known as a "living fossil tree." This divine tree is in a 29,333-hectare diversiform-leaved poplar forest in Ejina. Now, in the area of more than 30 meters surrounding this particular tree, five younger ones have grown up. Local herders call the group five "Mother and Children." The flourishing tree family stands majestically amid red willows in the sand dunes.

中文名：胡杨
拉丁名：*Populus euphratica* Oliv.
所在地：内蒙古自治区阿拉善盟额济纳旗
树龄：约 880 年
胸（地）围：850 厘米
树高：2700 厘米
冠幅（平均）：2095 厘米

Chinese Name: 胡杨
Latin Name: *Populus euphratica* Oliv.
Location: Ejina Banner, Alashan League, Inner Mongolia Autonomous Region
Tree Age: About 880 years
Chest (Floor) Circumference: 850 cm
Tree Height: 2,700 cm
Crown Width (Average): 2,095 cm

最美枫杨：神农架枫杨

位于湖北省神农架林区松柏镇八角庙村的这株 707 年树龄的枫杨，地围 8.88 米，树高 26.8 米，平均冠幅 35 米。

相传古时候，神农架林区松柏镇八角庙村有一地主，经常欺压雇工。一天，一名雇工太累，在树下睡着了，地主发现后，带着家丁要把他捆绑起来进行惩罚。家丁正要殴打雇工时，天空突然电闪雷鸣，将家丁刚刚举起的皮鞭击落，地主和家丁吓得灰溜溜地跑了。从此以后，当地人奉此树为神明，常有人来此祭拜，香火不断。

这棵枫杨树根部曾分为三枝，其中一枝已被雷击断，尚存两枝，北边一枝横向生出 24 米，已用水泥墩支撑。树干受蚁害已空心，东南向的根部曾被火烧。为确保这棵树得到永久性保护，神农架林区党委政府和林业主管部门将此树周围筑起了围墙，并在中间重新填充了肥沃的泥土，设立了国家一级保护古树界牌，以杜绝人为破坏。

正是由于这棵古老的枫杨树，当地成为神农架的一个重要景点。每逢周末，松柏城区的上班族，都会带着孩子前来游玩。

The Most Beautiful Chinese Ash: A Chinese Ash in Shennongjia Forest

Located in Bajiaomiao Village, Songbai Town nestled in the Shennongjia Forest Area of Hubei Province, this 707-year-old Chinese ash has a ground circumference of 8.88 meters, a tree height of 26.8 meters and an average crown width of 35 meters.

Legend has it that, in ancient times, there was a landlord in Bajiaomiao Village often bullied his tenants. One day, a tenant was very tired and fell asleep under a tree. The landlord found out and had him bound up for punishment. When one of his hired roughnecks was ordered to beat the tenant with a whip, the sky was split by lightning accompanied by thunder, hitting the whip, sending the the landlord and thugs into terrified retreat. Since then, the local people have worshipped the tree as a Tree God, and many come to worship.

From the root of this Chinese ash grow branch trees. One of the three has been broken by a lightning strike. The other two have survived, with the northern one stretching out 24 meters and supported with a concrete pier. The trunk has been hollowed out by ants, and its southeast root was once burned. In order to ensure the permanent protection of this tree, the management committee of the Shennongjia Forest Area and the competent forestry department had a protective wall built around it, the inside of which was filled with fertile soil. A national first-class ancient tree protection signboard was set up to prevent human destruction.

Because of this ancient Chinese ash, the area has become a major must-see spot in Shennongjia. At weekends, many locals come to enjoy the scenery.

中文名：枫杨	Chinese Name: 枫杨
拉丁名：*Pterocarya stenoptera* C.DC.	Latin Name: *Pterocarya stenoptera* C.DC.
所在地：湖北省神农架林区松柏镇八角庙村	Location: Bajiaomiao Village, Songbai Town, Shennongjia Forest Area, Hubei Province
树龄：707 年	Tree Age: 707 years
胸（地）围：888 厘米	Chest (Floor) Circumference: 888 cm
树高：2680 厘米	Tree Height: 2,680 cm
冠幅（平均）：3500 厘米	Crown Width (Average): 3,500 cm

最美白榆："兄弟榆"

此树位于河北省张家口市赤城县样田乡上马山村。上马山村有古榆树两株。传说很久以前，村庄的两兄弟上山砍柴时，哥哥被雷电劈中变成了榆树，弟弟为了寻找哥哥最后也变成了榆树，所以当地人称这两株榆树为"兄弟榆"。两树相距咫尺，情同伯仲，大者粗壮需八人合抱，小者亦需五六人合抱。古榆树历经沧桑，仍巍然耸立，枝繁叶茂，生机盎然，枝如虬龙，旁逸斜出，树形奇特美观。

由于古榆树枝叶繁茂，每当下雨的时候，分出的枝干因不堪负重而自然下垂。2011年，当地林业部门为这两株古榆树人工搭建了"树干"支撑，避免榆树枝干掉落。

由于古榆树的历史太过久远，当地人在树上供奉了"树神"，并将红绳系在树枝上，以祈福平安。

The Most Beautiful White-Bark Elm: "Brother Elms"

This tree is located in Shangmashan Village of Yangtian Township in Chicheng County, Zhangjiakou City, Hebei Province. The village used to have two such elms. It is said that, a long time ago, two brothers lived in the village. One day, they went up the mountain to cut firewood. The elder brother was struck by lightning and turned into an elm. The younger brother finally turned into an elm himself when he kept looking for his brother. The local people call the trees "Brother Elms." They are very close to each other. The big one needs eight people with extended arms to embrace the trunk, and the small one needs six. After enduring many vicissitudes, the ancient elms are still towering with luxuriant branches. These look like a dragon flying into the sky from the trunk.

The luxuriant branches and leaves of the ancient elms force the branches to naturally droop because they can't bear the load. In 2011, the local forestry department built "trunk supports" for the two ancient elms to avoid falling branches and trunks.

The local people used to worship the "Tree God" and hung auspicious red ropes on the branches praying for peace and happiness.

中文名：白榆
拉丁名：*Ulmus pumila* L.
所在地：河北省张家口市赤城县样田乡上马山村
树龄：650 年
胸（地）围：750 厘米
树高：2800 厘米
冠幅（平均）：2300 厘米

Chinese Name: 白榆
Latin Name: *Ulmus pumila* L.
Location: Shangmashan Village, Yangtian Township, Chicheng County, Zhangjiakou City, Hebei Province
Tree Age: 650 years
Chest (Floor) Circumference: 750 cm
Tree Height: 2,800 cm
Crown Width (Average): 2,300 cm

最美大果榆："夫妻树"

此树位于吉林省珲春市新安街道迎春社区的"古榆游园"中。据推测，该树生长于清朝顺治年间，传说当年有两株，公元1900年，八国联军之一的沙俄侵入我国东北，树木遭到了破坏，后又经"二战"日寇摧残，现仅存一株。

"古榆游园"除了可以休闲游乐外，这株古榆树也是一道记载着珲春市发展和改革的"风景线"，铭记着珲春的历史。古榆树树干粗壮，4个人才将其环抱。冬天，站在古榆树下抬头仰望，高高的树枝伸向云层，仿佛要探索宇宙的奥妙；春天，它枝繁叶茂，生机盎然，给人一种踏实向上的力量。古榆树的美不仅仅体现在苍劲上，还在于它极其独特的外形，它的主树干上分出朝南朝北的两个分树干，分树干生长的高度近乎相同，每当树叶都长出来的时候，两个树干的叶子触及到一起，越长越茂盛，好似恋人在紧紧地拥抱，所以当地老百姓都称之为"夫妻树"。

珲春市民对这株古榆树有着深厚的感情，人们为了祈福，在这株古榆树上"挂红"，一圈圈的红线寄托着人们对古榆树的喜爱，同时也希望从这株古榆树上汲取力量。珲春市绿委办及市园林办十分重视对古榆树的保护，为古榆树加设围栏、登记、挂牌、修整枝叶、除虫等，所在街道为了加强对古树的保护，还经常组织工作人员、辖区居民清理周边垃圾，确保其有一个良好的生长环境。

The Most Beautiful Ulmus Macrocarpa Hance: "Husband-Wife Tree"

This tree is located in the Ancient Elm Garden of Yingchun Community of Xin'an Street in Hunchun City of Jilin Province. It is said that there were two such trees here in the reign if Emperor Shunzhi of the Qing Dynasty (1633–1911). However, in 1900 when Tsarist Russian troops, one of the Eight Allied Forces, invaded Northeast China, the trees was damaged. After the Japanese invasion in World War II, only one remained.

The Ancient Elm Garden is place not only for leisure and recreation, but also a show of development during the reform and opening up period in Hunchun City. Its trunk is so thick that four people need to link arms to embrace it. When one stands below the tree, the top is often shrouded in clouds in winter, but provides great shade in spring and summer. The beauty of this ancient elm lies not only in its vigorous strength, but also in its extremely unique shape. Its main trunk has two branches extending south and north respectively, both of the same height, impressing people with the image of lovers hugging. This prompts the locals to call it the "Husband-Wife Tree."

Hunchun people often come to worship by hanging red threads onto the tree to pray for luck and to express their deep feelings. The municipal authorities have built a fence to ban close contact with the tree, and organize leaf pruning, and pest control. Local people are also mobilized to clean up the garbage accumulating in the tree area.

中文名：大果榆
拉丁名：*Ulmus macrocarpa* Hance
所在地：吉林省珲春市新安街道迎春社区
树龄：361 年
胸（地）围：146.5 厘米
树高：1430 厘米
冠幅（平均）：900 厘米

Chinese Name: 大果榆
Latin Name: *Ulmus macrocarpa* Hance
Location: Yingchun Community, Xin'an Street, Hunchun City, Jilin Province
Tree Age: 361 years
Chest (Floor) Circumference: 146.5 cm
Tree Height: 1,430 cm
Crown Width (Average): 900 cm

中文名：杉木
拉丁名：*Cunninghamia lanceolata* (Lamb.) Hook.
所在地：福建省宁德市蕉城区虎贝乡彭家村
树龄：1130 年
胸（地）围：829 厘米
树高：2030 厘米
冠幅（平均）：2210 厘米

Chinese Name: 杉木
Latin Name: *Cunninghamia lanceolata* (Lamb.) Hook.
Location: Pengjia Village, Hubei Township, Jiaocheng District, Ningde City, Fujian Province
Tree Age: 1,130 years
Chest (Floor) Circumference: 829 cm
Tree Height: 2,030 cm
Crown Width (Average): 2,210 cm

■ 最美杉木："伞树"

此树位于福建省宁德市蕉城区虎贝乡彭家村旁，其胸围 8.29 米，为福建省胸围最大的杉木。据记载，此株杉木种植于唐僖宗李儇光启年间（885 年—888 年），系彭氏祖先迁居至此时所植。奇特的是，这株古杉木的树枝皆向下生长，村民说此树为其祖先倒插种植。

据说此树树头原有一大洞，为雷击所致。1949 年以前，人们把桌子放进洞内，人可在内玩耍，后来洞口慢慢愈合，到 20 世纪 80 年代，洞口还有五六十厘米宽。2013 年，洞口仅余缝隙，有蜜蜂出入。

这棵树是彭家村历史的见证，历经朝代更迭，仍被完整保存下来，至今依然生长旺盛，每年可结果 50 千克，确属罕见。当地百姓把它誉为"神树"，因其树形呈伞状，又称其为"伞树"。彭氏家谱有诗云："枝繁叶茂历悠悠，伴祖肇迁有千秋。馨竹国史传铭志，伞树家声万古流。"

宁德市将其载入《宁德县志》，作为重点文物进行保护。因其为全省胸径最大的杉木，具有极高的保护价值，2013 年，福建省绿委、福建省林业厅将其列为重点古树名木进行保护管理。

（图片摄影 庄晨辉）

■ The Most Beautiful China Fir: "Umbrella Tree"

This ancient tree is located next to Pengjia Village, Hubei Township, in Jiaocheng District of Ningde City, Fujian Province. Its chest circumference is 8.29 meters, being the largest in the province. According to records, this China fir was planted in sometime between 885–888 in the Tang Dynasty when the ancestors of the Peng family moved and settled down. Strangely, the branches of this ancient China fir grow downward. Villagers thus say the tree was planted upside down.

It is said that there was a big hole caused by a lightning strike in the head of the tree. Before 1949, people moved a table into the hole to hold games. Strangely, the hole closed up slowly. By the 1980s, it was still 50–60 cm wide. by 2013, Only bees could fly in and out from the hole.

This tree is a witness of Pengjia Village's history. Despite dynastic changes, it still stands intact and bears 50 kg of fruit each year. No wonder the locals call it a "sacred tree." When viewed from afar, it resembles a giant umbrella, so it is called the "umbrella tree." A poem in *Peng's Genealogy* lauds it thus: "The branches are luxuriant and the history is long, it is 1,000 years old since the time of our ancestors...."

The tree has found its way into the *Annals of Ningde County* and is protected as a key cultural relic. This tree, with the largest DBH among the China firs in the province, it has since 2013 been listed as a key ancient tree enjoying protection in Fujian Province.

（Photos by Zhuang Chenhui）

最美柳杉："柳杉王"

此树位于浙江省丽水市景宁县大漈乡大漈村时思寺门前东侧斜坡上，至今已有1500年的历史。古树原高50多米，胸围13.4米，平均冠幅16米，要10多人手拉手才能合抱。经植物专家鉴定，它不仅是全国最大，也是亚洲最大的柳杉，《浙江古树名木》一书称它是柳杉世界里的"王中王"。2001年秋，这棵柳杉经过多方考证，被认为是世界上最大、最古老的柳杉树，并被称为"柳杉王"。

"柳杉王"历经千年的风雨侵袭、雷电轰击，顶部已折断，主干一半已枯，树身内空透天，唯有两边擎天"巨臂"依然青翠葱茏，生机勃勃。古树主干被雷击截断，削去大半，现仅有29米，奇特的是：其树干空心，根部有一个形似门户的洞，一人可自由进出，进到树洞中，抬头可见日光，如同坐井观天。树洞空间奇大，可摆一张圆桌供10余人围桌共餐。

大漈乡一位叫梅林的老人首先发起倡议，要求对千年"柳杉王"进行精心保护。2009年，景宁县林业局委托浙江农林大学对柳杉王进行专业保护，通过改善柳杉的立地条件、侧枝固定、防腐处理、空洞填充、增加围栏、病虫害防治等措施，让柳杉重焕了生机。

The Most Beautiful Cryptomeria Fortunei: "King of Cryptomeria Fortunei"

The tree is located on the east side of the gate to the Shisi Temple, Daji Village, Daji Township, Jingning She Autonomous County, Zhejiang Province. The 1,500-year-old tree has a height of more than 50 meters, a chest circumference of 13.4 meters, and an average crown width of 16 meters. It takes more than 10 people to embrace it hand inhand. According to the appraisal of plant experts, it is the largest Cryptomeria fortunei in China and Asia as well. In the autumn of 2001, the Cryptomeria fortunei in Daji Township was rated as the largest and oldest one of its kind in the world, and was called the "King of Cryptomeria Fortunei."

After thousands of years, the "King of Cryptomeria Fortunei" has broken its top and half of its trunk has withered become hollowed. Only its two "giant arms" on both sides are still green and vigorous. The bulk of the trunk was cut off by lightning to a height of 29 meters. What is strange is that the lower part of the remnant trunk is hollow inside and has a kind of gate for one to enter freely. The tree hole is spacious enough to hold a round table for more than 10 people to eat together.

An old man named Mei Lin in Daji was the first to propose special protection ofthe "Millennium King of Cryptomeria Fortunei." In 2009, the Jingning County Forestry Bureau entrusted the Zhejiang A&F University to carry out professional protection for Cryptomeria fortunei. By improving the site conditions of the ancient tree, fixing the lateral branches, antiseptic treatment, filling empty holes, increasing fences, pest control and other measures, Cryptomeria fortunei has been revitalized.

中文名：日本柳杉
拉丁名：(*Cryptomeria japonica* (*Thunb. ex L. f.*) D. Don
所在地：浙江省丽水市景宁县大漈乡大漈村
树龄：1500 年
胸（地）围：1340 厘米
树高：2900 厘米
冠幅（平均）：1600 厘米

Chinese Name: 日本柳杉
Latin Name: *Cryptomeria japonica* (*Thunb. ex L. f.*) D. Don
Location: Daji Village, Daji Township, Jingning County, Lishui City, Zhejiang Province
Tree Age: 1,500 years
Chest (Floor) Circumference: 1,340 cm
Tree Height: 2,900 cm
Crown Width (Average): 1,600 cm

最美南方红豆杉：松阳红豆杉

此树生长在浙江省丽水市松阳县玉岩镇大树后村，和该村渊源很深。据当地族谱记载，北宋年间，大批王族、官员、士绅涌向江南归隐山水，洪姓括苍太守后裔一路走来，见此地大树华盖、景色秀丽，又能遮蔽风雪，便择林后安家。千百年来，子孙繁衍，遂形成村庄，这就是现在的大树后村。

这棵红豆杉树经历代村民精心呵护至今，屹立于幽静山谷之中，虽经千年的风霜雨雪，依然枝繁叶茂，遗世而独立，每到秋日，硕果累累，红果绿叶，分外妖娆。由于古树历经千年，其主干芯材腐朽，形成空洞和部分树干缺口，两米多长的空洞里，能放八仙桌一张，成为该古树独一无二的特征。

村里还流传着很多与古树相关的故事。据说有一年，有户村民的一头黄牛从牛栏逃出，当时电闪雷鸣，狂风暴雨，失牛的村民心急如焚。农耕时代，一头牛可是一家农户的一半家产，他们发动村民四处寻找却不见踪影，正当人们一筹莫展之际，有个村民却意外发现丢失的牛躲在大树洞中安然休息，是这棵古树无意中保护了村民的命根子，从此，村民们对这棵古树有了更深的感情。1935年，粟裕、刘英率领红军挺进师驻扎到大树后村，以此为中心，建立起了浙西南革命根据地。

The Most Beautiful Chinese Yew: The Chinese Yew in the Songyang County

This tree is located in Dashuhou Village, Yuyan Town, Songyang County, Lishui City, Zhejiang Province. This ancient tree has a deep connection with the village. According to the local genealogy, during the Northern Song Dynasty (960–1279), a large number of royal families, officials and gentry flocked to area south of the Yangtze River to live in seclusion. The descendants of Kuocang prefect chief surnamed Hong came to love this place with the beautiful trees and the beautiful scenery. They chose the forest where they settled down. Their descendants have since multiplied into the Dashuhou Village.

This Chinese yew has been carefully cared for by the villagers for generations. Up to now, the ancient tree in the quiet valley is still full of branches and leaves. Every autumn, it is rich in green leaves and laden with red fruits. After thousands of years, the core material of the ancient tree decayed, forming holes and some trunk gaps. In the two-meter-long hole, an "eight immortal table" can be placed, which has become the unique feature of the ancient tree.

Many stories told are related to this ancient tree. One year, one of the villagers' cattle escaped from the cowshed when there was lightning and thunder. Many villagers hit out to look for it. Suddenly, the lost cattle were found hiding in the big tree hole which provided shelter for them. In 1935, the Red Army troops led by Generals Su Yu (1907–1984) and Liu Ying (1905–1942) stationed in Dashuhou Village where they established the revolutionary base in Southwest Zhejiang.

中文名：南方红豆杉
拉丁名：*Taxus chinensis* (Pilger) Rehd.var.*mairei* (Lemee et Levl.) Cheng et L.K.Fu
所在地：浙江省丽水市松阳县玉岩镇大树后村
树龄：1200 年
胸（地）围：880 厘米
树高：3000 厘米
冠幅（平均）：1200 厘米

Chinese Name: 南方红豆杉
Latin Name: *Taxus chinensis* (Pilger) Rehd.var.*mairei* (Lemee et Levl.) Cheng et L.K.Fu
Location: Dashuhou Village, Yuyan Town, Songyang County, Lishui City, Zhejiang Province
Tree Age: 1,200 years
Chest (Floor) Circumference: 880 cm
Tree Height: 3,000 cm
Crown Width (Average): 1,200 cm

最美东北红豆杉：东北红豆杉"树王"

在吉林省延边州汪清林业局荒沟林场 63 林班，有一片 30 多株的红豆杉古树群落，其中最大的一株树高 20.5 米，平均冠幅 8.6 米，树龄约 3000 年，是汪清林业局迄今为止发现的最大的一株东北红豆杉。

东北红豆杉又名紫杉，是第三纪孑遗的珍贵树种。汪清林业局所在地区属于典型的温带大陆性湿润山地季风气候，加之肥沃的棕色森林土壤以及茂密的森林，造就了东北红豆杉的适宜生境，在世界范围内亦属罕见。

在汪清林区流传着入山祭拜的习俗，由于此株东北红豆杉树龄大，历史悠久，当地人都尊称它为"树王"。每逢进山采摘活动时，人们便带来酒肉献于古树之前，并焚香祭拜，祈求平安顺利。它树形美丽，果实成熟期红绿相映的颜色令人陶醉，吸引游人前来观赏，具有较高的观赏价值和文化底蕴。

古树表皮呈红褐色，高耸入云，虬枝峥嵘，历经沧海桑田，朝代更迭，进入新千年，在汪清这片土地上焕发出了勃勃生机。2002 年 12 月，汪清林业局申请成立了汪清省级自然保护区，主要保护东北红豆杉资源及其赖以生存的针阔混交林生态系统。2011 年 3 月，汪清林业局向国务院申请将省级保护区晋升为国家级保护区，2013 年 6 月，经国务院批准，晋升为吉林汪清国家级自然保护区。保护区总面积 67434 公顷，是我国境内面积最大的以东北红豆杉为主要保护对象的国家级自然保护区。

中文名：东北红豆杉
拉丁名：*Taxus cuspidata* Sieb.et Zucc.
所在地：吉林省延边州汪清林业局荒沟林场
树龄：3000 年
胸（地）围：528.5 厘米
树高：2050 厘米
冠幅（平均）：860 厘米

Chinese Name: 东北红豆杉
Latin Name: *Taxus cuspidata Sieb.et Zucc.*
Location: Huanggou Forest Farm, Wangqing Forestry Bureau, Yanbian Prefecture, Jilin Province
Tree Age: 3,000 years
Chest (Floor) Circumference: 528.5 cm
Tree Height: 2,050 cm
Crown Width (Average): 860 cm

The Most Beautiful Taxus Cuspidata: "Tree King"

In the Huanggou Forest Farm of the Wangqing Forestry Bureau, Yanbian Prefecture, Jilin Province, there is a group of 30-odd Taxus cuspidata. The largest is one which is 20.5 meters high, with an average crown width of 8.6 meters and a tree age of about 3,000 years. It is the largest of kind found so far.

Taxus cuspidata, also known as Yew, is a rare tree species relict in the Tertiary period. The area where these trees are found belongs to the typical temperate continental humid mountain monsoon climate. The area boasts the fertile brown forest soil and a dense forest, which combine to make the suitable habitat for Taxus cuspidata, which is also rare in the world.

People in the Wangqing forest area has a custom of going to worship the 3,000-year-old Taxus cuspidata, the well-known "tree king." Wine and meat are offered as sacrifices to the ancient tree and incenses burnt in pray for peace and luck. The old tree has a beautiful shape, as well as high ornamental value and cultural heritage.Lots of visitors smitten by the red and green color of the fruit come to visit the tree.

The ancient trees are reddish brown in skin, towering into the clouds and with flourishing branches. The 1,000-year-old trees have entered into the new millennium, and are full of vitality in the land of Wangqing. In December 2002, the Wangqing Forestry Bureau applied for the establishment of the Wangqing Provincial Nature Reserve mainly for the protection of the Taxus cuspidata resources and its coniferous and broad-leaved mixed forest ecosystem. In March 2011, the Wangqing Forestry Bureau applied to the State Council to upgrade the provincial nature reserve to a national nature reserve. In June 2013, with the approval of the State Council, it was promoted to be the Jilin Wangqing National Nature Reserve. With a total area of 67,434 hectares, the reserve is the largest national nature reserve with Taxus cuspidata as the main protection object in China.

中文名：长苞铁杉
拉丁名：*Tsuga longibracteata* Cheng
所在地：福建省宁化县治平乡邓屋村
树龄：800 年
胸（地）围：558.9 厘米
树高：4580 厘米
冠幅（平均）：2200 厘米

Chinese Name: 长苞铁杉
Latin Name: *Tsuga longibracteata* Cheng
Location: Dengwu Village, Zhiping Township, Ninghua County, Fujian Province
Tree Age: 800 years
Chest (Floor) Circumference: 558.9 cm
Tree Height: 4,580 cm
Crown Width (Average): 2,200 cm

最美长苞铁杉：长苞铁杉王

这株长苞铁杉古树，位于福建省宁化县治平乡邓屋村召光自然村的屋背山。关于这棵高耸入云、伟岸挺拔的古树村里流传着一个传奇故事，体现了宁化客家人热爱树木、保护树木的优良传统美德。

相传南宋年间，有一姓廖的男子与他的妻子由宁化县淮土乡南迁至治平乡崇兴坊，过着自由自在的田园生活。可让他烦恼的是妻子一直都没能为他添个一男半女。有一天，他家喂养的9只鸡不见了，这可把他们急坏了。夜晚，他翻来覆去睡不好觉，蒙眬中有一位老态龙钟的人来到床前，右手拄着拐杖，左手拿着一树枝，对廖姓男子说："你家的鸡在一处叫'召光'的山上。你要去东方找到一种树苗，然后在鸡睡过的地方种上9棵树。只要你把这9棵树种活了，就把家迁至那里安居，保证能生儿育女，子孙兴旺。"老人说完就扔下树枝，飘然而去。廖姓男子一惊，醒了，原来是南柯一梦。他赶忙叫醒妻子，将梦境说了一遍，夫妻俩都觉得奇怪，半信半疑。这时天已蒙蒙亮，于是就起床，刚出门就看到地上有一树枝，跟梦里老人手中拿的一模一样，不远处又看到了梦里老人用草结的记号，顺着记号一路找去，约半个时辰，来到一个山岗上，果真发现了自家的鸡，9只鸡睡在不同的9个地方。他便在鸡睡过的地方做上记号，带着鸡回家了。回家后，他马上拿着工具带上老人留下的树枝，来到东面山上寻找相同的树，不一会儿，就找到了，不多不少正好是9棵同样的树苗，他把这9棵树苗移植到了鸡睡过的地方。春天来了，这些树苗长出新芽，都成活了，于是他又按老人的话，把家迁了过来，与树为邻，傍山而居。果然，冬天妻子就为他生了一个大胖小子，后来又一连生了8个孩子。从此，他们的孩子就在召光这个地方定居下来，繁衍生息，子孙兴旺。

明末崇祯年间，政局动荡，烽烟四起，各地恶霸占地为王，烧杀抢夺，民不聊生。这时汀州有一恶霸想建一座豪宅大院，需要大量的木料，特别是需要8根粗大的木材做柱子。恶霸派出爪牙四处搜寻，终于在召光发现了他需要的长苞铁杉，于是恶霸亲自带着爪牙来到召光抢夺。全村上下奋勇阻拦，但都被打得遍体鳞伤，最后被抢走了其中的8棵长苞铁杉，余下的1棵便是存活至今的长苞铁杉王。

（图片摄影 庄晨辉）

The Most Beautiful Tsuga Longibracteata: Tsuga Longibracteata King

This old hemlock is located in Wubei Mountain, Zhaoguang Natural Village, Dengwu Village, Zhiping She Ethnic Township, Ninghua County, Fujian Province. There is a legend about this towering ancient tree, which reflects the fine traditional virtues of Ninghua Hakka people in loving and protecting trees.

It is said that in the Southern Song Dynasty (960–1279), a man surnamed Liao and his wife moved southward from Huaitu Township in Ninghua County to Chongxingfang in Zhiping Township, where they led an idyllic life. But what bothered them was that they had no children. One day, the nine chickens they had been feeding disappeared all of a sudden. At night, Liao had a dream: An old man in went over to him, with a crutch in his right hand a tree branch in his left hand. He said to Liao: "Your chickens are on the Zhaoguang Mountain. Finding nine tree branches in the place where your chicken rest and growing them into trees, and then move your home there. Then you can have many children." The old man left leaving behind nine tree branches. He woke up his wife and told the dream to her. Losing no time, the couple left for the mountain where they found a tree branch similar to the one in the hand of the old man. Not far from the tree branch, they found their nine chickens which rested in different places. They marked these places and then took the chicken back home. When he returned home, Liao went to the mountain in the east and found nine similar tree seedlings. He moved them to the place his chicken rested. When spring came, the trees sprouted new shoots and all survived. So, he moved his home again according to the old man's instructions, living next to the trees. Sure enough, his wife gave birth to a big fat boy for him in winter, and then gave birth to eight more children in a row. From then on, their children settled down in Zhaoguang, where they multiplied and prospered.

During the reign of Emperor Chongzhen at the end of the Ming Dynasty (1368–1644), the country was in turmoil and the flames of war were everywhere. The local tyrants occupied land and burned, killed and robbed the people who were in dire straits. At this time, there was a bully in Tingzhou who wanted to build a mansion. He was in urgent need of wood, especially eight thick wood pillars. The bully sent his minions to search around, who finally found the Tsuga Longibracteata he needed in Zhaoguang. The bully came to Zhaoguang with his paws. The whole village tried to stop them, but they were beaten black and blue. Finally, eight of them were robbed, the remaining one surviving as the "King of Tsuga Longibracteata" as we see today.

(Photos by Zhuang Chenhui)

最美油杉："谊父树"

该树位于福建省永泰县同安镇芹草村水口的芹草宫前，树姿舒展优美。据村民介绍，自明永乐三年陈氏先祖迁进芹草洋始盖祖屋时，长在水口处的这株油杉已是参天大树了。为使祖屋风水更具灵气，陈氏先祖将这棵油杉当作镇水口和护山门的树神，并在树旁建了一座小庙（即芹草宫），老油杉便成为了护庙树。

陈氏家族在大油杉的庇佑下不断繁衍壮大。村里人将宫前的这株油杉敬为神树，每年正月闹元宵的第一串鞭炮总是从油杉树下开响。村民还祈佑大油杉为他们祛病免灾。村中大人、小儿体弱多病的，多拜此树为谊父。谊拜选在黄道吉日，敬上礼品后三叩首，并在树身围系一根红绳见证。现此树已收有24个谊子。据村民介绍，村里有一个叫陈积珍的人，家住油杉树下，他13岁那年的一天下午，在油杉树旁放羊时，突遇山洪暴发被卷走。村里人闻讯赶到，沿溪流寻找，未果，大家都以为他已生还无望。谁知过了几个钟头，山洪退去，他毫发无损地回到家里。村里人都说是谊父树在庇护他。

（图片摄影 庄晨辉）

The Most Beautiful Chinese Fir: "Adoptive Father Tree"

This graceful tree is located in front of Qincao Temple, near the water gap of Qincao Village, in Tong'an Town of Yongtai County, Fujian Province. Villagers say that the tree was quite big when Chen's ancestors moved to Qincaoyang in the third year of the reign of Emperor Yongle of the Ming Dynasty (1368–1644) and established the ancestral house. In order to strengthen the *fengshui* of the ancestral house, Chen's ancestors took this Chinese fir as the "Tree God," and built the Qincao Temple beside it.

With the protection of the Chinese fir, the Chen's family flourished. Showing full respect for the "Tree God," People in the village still revere the Chinese fir tree in front of the temple as a sacred tree. When the Lantern Festival comes in the middle of the first lunar month, firecrackers are set off under the tree in pray for good health and good luck. The villagers, especially those with weak physique, revered the tree as their Adoptive Father. For this purpose, they chose an auspicious day and came to kowtow and offer burning incense, and hung a red rope on the branches. One time, a 13-year-old boy named Chen Jizhen, whose home located under the Chinese fir tree, was washed away by a mountain flood when he was out herding, and was nowhere to be found. When the mountain flood receded, he returned home unharmed. Everyone in the village said it was the Adoptive Father Tree who had protected him.

（Photos by Zhuang Chenhui）

中文名：油杉
拉丁名：*Keteleeria fortunei* (Murr.) Carr.
所在地：福建省永泰县同安镇芹草村
树龄：1500 年
胸（地）围：672 厘米
树高：2580 厘米
冠幅（平均）：2870 厘米

Chinese Name: 油杉
Latin Name: *Keteleeria fortunei* (Murr.) Carr.
Location: Qincao Village, Tong'an Town, Yongtai County, Fujian Province
Tree Age: 1,500 years
Chest (Floor) Circumference: 672 cm
Tree Height: 2,580 cm
Crown Width (Average): 2,870 cm

最美铁坚油杉："夫妻树"

在大巴山深处，重庆与陕西交界的重庆市巫溪县鱼鳞乡五宋村4组，有一座道教宫观——五龙宫，两棵千年铁坚油杉古树在这里并列而生，被当地人称为"夫妻树"。这里香火鼎盛，特别是每年正月初一、十五，各地香客慕名而来，虔诚拜谒。

"夫妻树"单株胸围约2.8米，树高35米，冠幅29.5米，树龄1570年左右。国家和重庆市相关林业、植物专家和美国学者曾来此进行考证，据说是亚洲目前发现的最大的铁坚油杉树。

唐朝时期，奉节曾为夔府，掌管周边多县（包括现今的巫溪）。府官每日晨起洗脸时，铜盆中总是显现两棵大树，甚感怪异。于是派人四处查找，历经千辛万苦，终于在大山深处的五龙宫发现了这两棵树，与盆中所见一模一样。两株"神树"恰好地处盐马古道要冲，于是府官在此设衙，署理事务，久而久之，此地成为周边政治经济文化中心，盛极一时。

现代科学考证，铁坚油杉雌雄同株，每棵树都能结果。而五龙宫这两棵树，人们一直相传为一公一母，"公"树不结果，"母"树结果，因此被称为"夫妻树"。当地百姓不生孩子就去求其赐子，找不到媳妇就去求其赐个媳妇……在人们眼里，它们就是消灾祛病、求子问吉的"神树"。

近年来，很多人慕名而来，烧香拜谒。"神树"自此声名远扬。

The Most Beautiful *Keteleeria Davidiana*: "Husband and Wife Trees"

In the depths of the Daba Mountain in Group Four of Wusong Village in Yulin Township, Wuxi County of Chongqing City, on the border of Chongqing City and Shaanxi Province, there is a Taoist temple called Wulong Temple where there are two 1,000-year-old *Keteleeria Davidiana* trees, known respectively as the Husband Tree and Wife Tree. Incense is very popular here, especially on the first and 15th day of the first lunar month every year.

Each of the trees has the DBH (Diameter at Breast Height) of about 2.8 meters, a height of 35 meters and a crown width of 29.5 meters. Their age has been established as 1,570 years. According to forestry and plant experts from Chongqing and American scholars, they are the largest *Keteleeria Davidiana* fir trees found in Asia.

In the Tang Dynasty (618–907), Fengjie County was once the site of prefectural government of Kuizhou, governing several surrounding counties (including today's Wuxi County). The government head found every morning when he washed his face with a basin, he would always see reflections of two strange trees in the basin. He dispatched his men to look for trees matching his description in the mountain and they eventually found the two trees on an ancient salt transportation road hub in the deep mountain. A government office was hence set up there to handle government affairs, since when the place was developing into a political, economic and cultural center in the area.

According to modern scientific research, *Keteleeria fortunei* is monoecious, and each tree can bear fruit. The two trees in the Wulong Temple have been identified one male and one female. The male tree does not bear fruit, but the female does, resulting in the name of "Husband and Wife Trees." If local people have no children, they turn to the trees. For men without a wife, they again pray to the trees to find one. In the eyes of the locals, they are holy trees.

In recent years, many people come to burn incense and the "holy trees" have become widely known.

中文名：铁坚油杉
拉丁名：*Keteleeria davidiana* (Bertr.) Beissn.
所在地：重庆市巫溪县鱼鳞乡五宋村
树龄：1570 年
胸（地）围：880 厘米
树高：3500 厘米
冠幅（平均）：2950 厘米

Chinese Name: 铁坚油杉
Latin Name: *Keteleeria davidiana* (Bertr.) Beissn.
Location: Wusong Village, Yulin Township, Wuxi County, Chongqing City
Tree Age: 1,570 years
Chest (Floor) Circumference: 880 cm
Tree Height: 3,500 cm
Crown Width (Average): 2,950 cm

最美水杉：利川"水杉王"

利川"水杉王"位于湖北省利川市谋道镇水杉植物园。它是世界上树龄最大、胸径最大的水杉母树，并被世界公认为"天下第一杉""水杉王"，对古植物、古气候、古地理和地质学，以及裸子植物系统发育的研究均有重要的意义。

目前，全球存活的大部分水杉树，都是这棵"水杉王"繁殖的后代。因此，它又被称为全世界水杉的"母树"。"水杉王"被发现之后，植物学家又陆续在利川谋道至小河一带，找到了5700多棵水杉树。

水杉树树干通直挺拔，树枝向侧面斜伸出去，全树犹如一座宝塔。其枝叶扶疏，树形秀丽，既古朴典雅，又肃穆端庄，树皮呈赤褐色，叶子细长，很扁，向下垂着，入秋以后便脱落。

1943年，植物学家王战教授在四川万县磨刀溪路旁（现谋道镇）发现了一棵高达33米，树围2米的奇异树木。当时谁也不认识它，甚至不知道它应该属于哪一属、哪一科。一直到1946年，由我国著名植物分类学家胡先骕和树木学家郑万钧共同研究，才证实它就是亿万年前在地球大陆生存过的水杉，从此，植物分类学中就单独添进了一个水杉属、水杉种，引起世界轰动。

The Most Beautiful Chinese Redwood: "Chinese Redwood King" in Lichuan

The tree is located in the Chinese Redwood Botanical Garden in Moudao Town in Lichuan City of Hubei Province. It is the mother tree of Chinese redwood in China and the first Chinese Redwood in the world. It is of great significance to the study of palaeoflora, paleoclimate, palaeogeography and geology, as well as the phylogeny of gymnosperms.

At present, most of the surviving Chinese Redwoods in the world are the offspring of this Chinese Redwood King. Therefore, it is also known as the "mother tree" of Chinese Redwoods in the world. After the discovery of this tree, botanists found more than 5,700 Chinese redwoods spread across the area from Moudao Town of Lichuan to Xiaohe.

The tree trunk is straight, and the branches are inclined out. The whole tree looks like a pagoda. Its branches and leaves are sparse; the tree shape is beautiful, simple and elegant, as well as solemn and dignified; the bark is reddish brown; the leaves are long and thin, very flat, pointing downward, and are shed in late autumn.

In 1943, Professor Wang Zhan, a botanist, found a strange tree with a height of 33 meters and a circumference of two meters near Modaoxi Road (now Moudao Township) in Wanxian County, Sichuan Province. At that time, no one knows anything about it, not even what genus or family to which it should be ascribed. It was not until 1946, when Hu Xiansu, a famous plant taxonomist in China, and Zheng Wanjun, a dendrologist, jointly studied the tree, and found that it was a kind of dawn redwood that had emerged hundreds of millions of years ago. Since then, a genus and species of Chinese redwood have been added to the taxonomy of plants, regarded as a world-shaking event.

中文名：水杉
拉丁名：*Metasequoia glyptostroboides* Hu et Cheng
所在地：湖北省利川市谋道镇水杉植物园
树龄：850 年
胸（地）围：778.72 厘米
树高：3500 厘米
冠幅（平均）：2200 厘米

Chinese Name: 水杉
Latin Name: *Metasequoia glyptostroboides* Hu et Cheng
Location: Redwood Botanical Garden, Moudao Town, Lichuan City, Hubei Province
Tree Age: 850 years
Chest (Floor) Circumference: 778.72 cm
Tree Height: 3,500 cm
Crown Width (Average): 2,200 cm

最美侧柏：黄帝手植柏

这株举世闻名的古柏，生长在海拔 1050 米的陕西省延安市黄陵县轩辕庙院内，高 19.5 米，胸围 8.38 米，平均冠幅 18 米。相传它为轩辕黄帝亲手所植，故名为黄帝手植柏，距今 5000 多年。其树枝像虬龙在空中盘绕，苍劲挺拔、冠盖蔽空、层层密密，像个巨大的绿伞。

1982 年，英国林业专家罗皮尔考察了 27 个国家的柏树后，认为唯有黄帝手植柏最粗壮、最古老。"中华名树公选养护委员会"将它评为"中华百棵名树之首"。世人誉之为"世界柏树之冠"。当地谚语这样描绘它的粗壮："七搂八拃半，圪里圪瘩不上算。"就是说，7 个人手拉着手合抱不拢树干，还剩 8 拃多。"黄帝手植柏"沐浴了 5000 年的风风雨雨，目睹了中华民族的荣辱兴衰，至今依然苍翠挺拔、枝繁叶茂，彰显出中华民族生生不息、国脉传承的强大生命力。

The Most Beautiful Chinese Arborvitae: Cypress Planted by the Yellow Emperor

This world-famous ancient cypress grows in the Xuanyuan Temple courtyard in Huangling County of Yan'an City, Shaanxi Province, at an altitude of 1,050 meters. It is 19.5 meters high, and has a chest circumference of 8.38 meters and an average crown width of 18 meters. It is said to have been planted by the legendary Yellow Emperor Xuanyuan (2697–2599 BCE). Its branches are like a winding dragon in the air, and its canopy is like a huge green umbrella.

In 1982, British forestry expert Robert Roper inspected the cypress trees in 27 countries and concluded that this particular cypress was the strongest and oldest. The Chinese Famous Trees Public Election and Conservation Committee rated it as "the top of 100 Famous Chinese Trees." It is known as the "Crowning Glory of Cypress Trees in the World." It is so thick that even seven people holding hands cannot embrace the trunk. In the past 5,000 years, the tree witnessed the rises and falls of the Chinese nation. It is still standing firmly and stretching luxuriant branches, embodying the tenacity of the lives of Chinese nation.

中文名：侧柏
拉丁名：*Platycladus orientalis* (L.) Franco
所在地：陕西省延安市黄陵县轩辕庙
树龄：5000 年
胸（地）围：838 厘米
树高：1950 厘米
冠幅（平均）：1800 厘米

Chinese Name: 侧柏
Latin Name: *Platycladus orientalis* (L.) Franco
Location: Xuanyuan Temple, Huangling County, Yan'an City, Shaanxi Province
Tree Age: 5,000 years
Chest (Floor) Circumference: 838 cm
Tree Height: 1,950cm
Crown Width (Average): 1,800 cm

中文名：圆柏
拉丁名：*Sabina chinensis* (L.) Ait.
所在地：河北省邯郸市临漳县习文乡靳彭城村
树龄：1500 年
胸（地）围：550 厘米
树高：2100 厘米
冠幅（平均）：1733 厘米

Chinese Name: 圆柏
Latin Name: *Sabina chinensis* (L.) Ait.
Location: Jinpengcheng Village, Xiwen Township, Linzhang County, Handan City, Hebei Province
Tree Age: 1,500 years
Chest (Floor) Circumference: 550 cm
Tree Height: 2,100 cm
Crown Width (Average): 1,733 cm

最美圆柏：曹魏古柏

此树位于河北省邯郸市临漳县习文乡靳彭城村，即全国重点文物保护单位——邺城遗址曹魏南校场处，东有魏齐寺院，南有玄武池、籍田、招贤馆旧址，西有魏文帝甄后朝阳陵。古树树身伟岸，虬瘤突兀，枝繁叶茂，郁郁葱葱，饱经沧桑而未毁损，久历岁月而不衰，在古来战乱频起的中原地区是一个奇迹。

相传三国时期，曹操为了打过长江，一统天下，在铜雀台南面兴修了南校场和玄武池。曹操在举行籍田、阅兵、训水师仪式时找不到拴马之处，次子曹植见状，特意从太行山移来一棵碗口粗细的柏树，种在玄武池南。柏树汲取漳南大地的灵气，越发长得挺拔茂盛，曹操见状非常高兴，每当骑马到此，总把马拴在这棵树上，因此，这棵柏树也就有了"曹操拴马桩"的美称。

三国归晋，沧桑巨变。北齐时，曹魏古柏处建为彭城，彭城王居住此地，古柏院落成为了彭城佛教圣地，一方净土。唐宋时，曹魏古柏处发展为道、儒、佛教圣地，礼化百姓，故曹魏古柏又称为彭城三教堂大柏树。元明清三代，风霜剑雨，电闪雷击。曹魏古柏依然枝繁叶茂，苍翠挺拔，当地百姓以为古柏已升仙有灵，能庇护四方，可尚农安居乐业，临战无伤死战亡，遂供奉为"柏仙"，又称群仙居此树冠。

历经千年锤炼，如今，这棵大柏树依然枝叶繁茂，长势独特。从不同方向看那枝杈，古柏有许多神奇的形态。

The Most Beautiful Sabina: Beautiful Ancient Sabina of the Wei State of the Three-Kingdoms Period

The ancient tree stands in Nanxiaochang of the Yecheng Wei State Ruins of Jinpengcheng Village, Xiwen Township, Linzhang County in Hebei Province. It has the Weiqi Temple lying in the east, and Xuanwuchi, Jitian and the Site of Zhaoxian Hall in the south, as well as the Chaoyang Mausoleum of Empress Zhen of the Wei State in the south. The ancient tree has a great body, protruding burrs, and luxuriant branches. It has gone through many vicissitudes without being seriously damaged, thus regarded as a miracle.

It is said that, during the Three Kingdoms period (220–265), in order to fight his way across the Yangtze River and unify the whole country, Cao Cao built Nanxiaochang and Xuanwuchi in the south of Tongquetai. When Cao Cao organized farming, and conducted military parades and training there, he was unable to find a place to tie his horse. His second son, Cao Zhi, especially moved a cypress from Taihang Mountain and planted it in a place south of Xuanwuchi. The cypress later became known as the "Horse Fastening Pillar of Cao Cao."

The Three Kingdoms fell and the Jin Dynasty (266–420) rose. In the Northern Qi Dynasty (266–420), Pengcheng Town was built where this ancient cypress was and became home to the ruler of Pengcheng. The courtyard where the ancient tree stands became the Buddhist site. In the Tang and Song dynasties (960–1279), the ancient tree courtyard was holy to Taoism, Confucianism and Buddhism witnessing rituals of all three. In the period of the Yuan, Ming and Qing dynasties (1271–1911), the ancient tree was still luxuriant and vigorous. Local people thought that it had become immortal offering protection for locals from disaster and wars and becoming an object of worship.

Looking at the luxuriant branches from different directions, the ancient tree has many magical forms.

最美大果圆柏：热振森林公园大果圆柏

大果圆柏位于西藏自治区拉萨市唐古乡热振森林公园内。作为西藏中北部原始森林边缘地区最具特色的高山林灌植被，大果圆柏具有很高的科研价值和观赏价值。

同时，大果圆柏所在地有着丰富的人文景观资源，藏传佛教噶当派的第一座寺庙——热振寺位于园内。因此，这里成为了极具吸引力的朝拜圣地，为开展生态旅游创造了良好的基础条件。

The Most Beautiful China Savin: Big-Fruit Tree in Razheng Forest Park

The big-fruit China Savin tree is located in the Razheng National Forest Park in Tanggu Township, Lhasa City, Tibet Autonomous Region. As the most characteristic high mountain forest and shrub vegetation at the edge of primitive forest in Central and Northern Tibet, the tree has high scientific research and ornamental value.

This ancient tree is located in a place that holds abundant cultural sites, for example, Razheng Monastery, the first monastery of the Gagyu Sect of Tibetan Buddhism. With reputation as a holy site for worshipers, it now serves as a boon for the development of eco-tourism.

中文名：大果圆柏
拉丁名：*Sabina tibetica* Kom
所在地：西藏自治区拉萨市唐古乡
树龄：500 年
胸（地）围：503 厘米
树高：1000 厘米
冠幅（平均）：540 厘米

Chinese Name: 大果圆柏
Latin Name: *Sabina tibetica* Kom
Location: Tanggu Village, Tanggu Township, Lhasa City, Tibet Autonomous Region
Tree Age: 500 years
Chest (Floor) Circumference: 503 cm
Tree Height: 1,000 cm
Crown Width (Average): 540 cm

最美国槐："天下第一槐"

此树位于河北省邯郸市涉县固新镇固新村，有"固新老槐树，九搂一屁股"之说，是目前我国已知的树龄最长的槐树，又有"天下第一槐"之美誉。

古槐历经风霜雨雪仍傲然挺立，夏可遮阴避暑，冬可挡风拒寒，曾解救人饥寒，其高贵品格，确实值得后人咏颂。据嘉庆四年《涉县志》记载："树大十数围，枝叶扶疏，状类虬龙"。另据该村《古槐碑记》记载："中州胜地古槐者源溯沙候国（即涉县）属地也，周十数围，高入云霄，世人罕见。乃中华灵秀之种，民族之骄也。槐寿几何有待于考。但有民间佳话盛传：一曰大明正德初叶立村之时已有古槐千年之说；二曰战国时期，秦兵攻赵东进路过于此曾歇马饮食；三曰唐代吕翁在此修道，德高好弈，有先天古槐后世小仙之语；四曰其槐枝繁叶茂，延伸四方，覆盖数亩，曾有'槐荫福地'盛誉匾额高悬；五曰明末灾荒古槐开仓以槐豆树叶拯救饥民，昼采夜长，茂然不败。历史久远，惜古槐遭自然侵袭与异族摧折，故失原貌……"

古槐是清漳河谷千百年文明史的见证。随着历史变迁，日久天长，古槐原有树冠主枝均已枯朽，但萌生的新枝年年花繁叶茂，展示着顽强的生命力，令人称奇。

The Most Beautiful Scholar Tree: "The Best in the World"

This tree is located in Guxin Village, Guxin Township, Shexian County of Handan City in Hebei Province. It is known as the oldest tree of its kind in China.

The ancient tree offers shade to the locals in summer and shelter on rainy days. According to the *Annals of Shexian County* published in 1799, the fourth year of the reign of Qing Emperor Jiaqing, its girth was large enough to require 10 people to hold by hands. According to the record of the tree inscribed on the village's stone tablet, such a tree is rarely found. Some say it is 1,000 years old; some say that during the Warring States period (475–221 BCE), Qin State once sent troops to attack the State of Zhao, and the troops camped the tree. In the Tang Dynasty, it's said Lǚ Dongbin, a famous Taoist, had been cultivating himself according to Taoist doctrines in the area; at the time, the tree had luxuriant branches and leaves, and a horizontal board was erected inscribed with words meaning the tree brought blessings to the area; At the end of the Ming Dynasty, people suffered from drought and the government offered food made of beans and leaves of the tree to the starving people, and the tree remained sturdy and strong. However, it lost its original shape when it suffered from pest invasion and other damages.

This ancient tree is the witness to thousands of years of civilization in the Qingzhang River Valley. Although the old part of the tree has become withered and decayed, new branches continued to sprout year by year, showing a strong vitality which amazed all.

中文名：国槐
拉丁名：*Sophora japonica* Linn.
所在地：河北省邯郸市涉县固新镇固新村
树龄：2500 年
胸（地）围：1700 厘米
树高：2000 厘米
冠幅（平均）：1250 厘米

Chinese Name: 国槐
Latin Name: *Sophora japonica* Linn.
Location: Guxin Village, Guxing Township, Shexian County of Handan City in Hebei Province.
Tree Age: 2,500 years
Chest (Floor) Circumference: 1,700 cm
Tree Height: 2,000 cm
Crown Width (Average): 1,250 cm

中国最美古树
The Most Beautiful Ancient Trees in China

中文名：樟
拉丁名：*Cinnamomum camphora* (L.) Presl
所在地：福建省德化县美湖乡小湖村
树龄：1300 年
胸（地）围：1672 厘米
树高：2550 厘米
冠幅（平均）：3740 厘米

Chinese Name: 樟
Latin Name: *Cinnamomum camphora* (L) Presl
Location: Xiaohu Village, Meihu Township, Dehua County, Fujian Province
Tree Age: 1,300 years
Chest (Floor) Circumference: 1,672 cm
Tree Height: 2,550 cm
Crown Width (Average): 3,740 cm

最美樟树：德化"樟树王"

福建省德化县美湖乡小湖村有一株古樟树，虽历经千余年的风霜，仍茁壮挺拔，枝繁叶茂，如擎天大伞，庇荫人间大地。

据当地老人介绍，这株古樟的树干内曾腐朽成一个大洞，洞里能摆放一张方桌，如今洞已经看不见了，原来年年生长的新生皮层不仅将洞口密封起来，而且把立在树下的一块墓道碑的基部也包裹了三分之一，可见这棵古樟生命力多么顽强。

据《德化县志》记载，这株古樟植于唐代。据传在唐末五代时，有章姓和林姓二人为避黄巢起义战乱到了小湖村。两人筋疲力尽，躺在樟树下，倦极而眠，梦见一身披树叶的老翁站在面前，对他们说了四句隐语："两氏与吾本同宗，巧遇机缘会一堂，来年同登龙虎榜，衣锦荣归济四乡。"他俩醒来时身边只见这棵樟树，便恍然大悟，老翁所指"同宗"不正是林字的"木"旁加上"章"字成"樟"吗？于是，他们便在这樟树旁边建屋定居，日夜苦读，后来果然双双高中，衣锦还乡，为乡民办了许多好事，名垂青史。

当地村民为纪念他们，便在樟树旁建起了一座"章公庙"，又称"小龙庙"或"显应庙"，每年农历三月十六，村民们都会举办盛大的祭樟树王活动。

（图片摄影 张圆圆）

The Most Beautiful Camphor Tree: "King of Camphor Trees" in Dehua

A 1,000-year-old camphor tree in Xiaohu Village of Meihu Township, Dehua County, Fujian Province, still looks like a giant umbrella holding up to the sky to shade the earth.

According to locals, the trunk of this ancient tree once decayed into a big hole large enough for the insertion of a square table. Now, however, the hole has become invisible. The original annual growth of neocortex not only sealed the hole, but also wrapped the base of a tombstone standing under the tree by one-third, which shows how tenacious the ancient tree is.

According to the *Annals of Dehua County*, this ancient camphor was planted in the Tang Dynasty. It is said that, in the Five Dynasties (907–960), two men with the surnames of Zhang and Lin fled from the Huangchao Uprising chaos to Xiaohu Village. They were exhausted and lay under the camphor tree and soon fell asleep. They dreamt of an old man wrapped in leaves was standing in front of them, informing that they and he were from the same clan who had come together by chance; they would succeed in the imperial examination together in the coming year and return in glory. When they awoke, they saw this camphor tree, and decided to build homes and settle down nearby. They studied hard and passed imperial examinations and returned to do many good things for the locals.

In honor of them, a temple was built next to the tree which is called Mr. Zhang Temple, Little Dragon Temple and Xianying Temple. On the 16th day of the third lunar month, the villagers hold a grand ceremony to worship the tree king.

(Photos by Zhang Yuanyuan)

最美榕树:"小鸟天堂"

此树位于广东省江门市新会区会城街道天马村,小鸟天堂古榕树,原名"雀墩"。

390多年前,由河中一个泥墩中的一棵榕树长期繁衍,形成婆娑的榕叶笼罩着20多亩的河面,榕树枝干上长着美髯般的气生根,着地后木质化,抽枝发叶,长成新枝干。新干上又长成新气生根,生生不已,变成一片根枝错综的榕树丛,形成独木成林的奇观。更令人称奇的是,这棵神奇的古榕树上栖息着数以万计的各种野生鹭鸟,其中以白鹭和灰鹭最多。白鹭朝出晚归,灰鹭暮出晨归,一早一晚,相互交替,盘旋飞翔,嘎嘎而鸣,是世间罕有的"百鸟出巢,百鸟归巢"奇特景观。

1933年,文学大师巴金先生乘船游览后叹为观止,写下优美散文《鸟的天堂》。1958年,时任广东省委第一书记的陶铸与时任广东省省长陈郁视察新会区期间,观赏了当时仍叫"雀墩"的古榕,陶铸建议,根据巴金先生《鸟的天堂》的文意,改为"小鸟天堂"。为此,新会区政府部门专门约请巴老并获欣允,巴老先后于1982年和1984年分别亲笔书写了"小鸟天堂"的题名。1978年人民教育出版社把巴金《鸟的天堂》列入了全国小学六年级下学期《语文》教科书。

The Most Beautiful Banyan Tree: "Birds Paradise"

This is located in Tianma Village, Huicheng Street of Xinhui District, Jiangmen City in Guangdong Province. The ancient Banyan tree in a paradise for birds is formerly known as "Sparrow Pier."

More than 390 years ago, a Banyan tree was discovered in a muddy riverbed, with a crown spreading to cover more than one hectare of the river. The tree branches had beautiful bearded aerial roots which became lignified and sprouted their own branches to become a Banyan cluster with intricate roots and branches, forming a marvelous forest from a single trees. Surprisingly, there are tens of thousands of wild herons perching on this magical ancient Banyan tree, among which egrets and grey herons are the most common. Egrets come out in the morning and return in the evening, while gray herons return in the morning and night, alternately, hovering and creaking. It is a rare spectacle even in the world that "hundreds of birds come out of their nests with another hundreds returning."

In 1933, Ba Jin (1904–2005), a literary master, wrote a beautiful prose entitled *Birds Paradise* after a boat tour there. In 1958, Tao Zhu, then First Secretary of the CPC Guangdong Provincial Committee, and Chen Yu, then Governor of Guangdong Province, visited the "Sparrow Pier" during a special tour. They suggested that the "Sparrow Pier" be changed to "Birds Paradise" according to Ba Jin's work. Xinhui County government especially invited Ba Jin to inscribe the title of *Birds Paradise* in 1982 and 1984 respectively. In 1978, the People's Education Press listed Ba Jin's work into the textbook for the second semester of the sixth grade of primary schools in China, thus making the county known to all.

中文名：榕树
拉丁名：*Ficus microcarpa* Linn.
所在地：广东省江门市新会区会城街道天马村
树龄：394 年
胸（地）围：隔水，不宜测量
树高：1500 厘米
冠幅（平均）：15900 厘米

Chinese Name: 榕树
Latin Name: *Ficus microcarpa* Linn.
Location: Tianma Village, Huicheng Street, Xinhui District, Jiangmen City, Guangdong Province
Tree Age: 394 years
Chest (Floor) Circumference: not suitable for measurement because of water
Tree Height: 1,500 cm
Crown Width (Average): 15,900 cm

中文名：绿黄葛树
拉丁名：*Ficus virens* Ait.
所在地：贵州省安顺市关岭布依族苗族自治县岗乌镇中兴村
树龄：1000 年以上
胸（地）围：1735 厘米
树高：3000 厘米
冠幅（平均）：4400 厘米

Chinese Name: 绿黄葛树
Latin Name: *Ficus Virens* Ait.
Location: Zhongxing Village, Gangwu Town, GuanLing Autonomous County, Anshun City, Guizhou Province
Tree Age: over 1,000 years
Chest (Floor) Circumference: 1,735 cm
Tree Height: 3,000 cm
Crown Width (Average): 4,400 cm

最美绿黄葛树:"上甲贵古榕树群"

贵州省安顺市关岭布依族苗族自治县上甲布依古寨旅游景区,"上甲贵古榕树群"闻名于全市。该地古榕树众多,主要以绿黄葛树为主,年龄达千年的有数株。其中最独特的一株绿黄葛树,茎干粗壮,树形奇特,悬根露爪,蜿蜒交错,古态盎然;树叶茂密,叶片油绿光亮;枝杈密集,大枝横伸,小枝斜出虬曲,游人见之无不惊叹。

近年来,全国各地乃至海外的林业机构慕名而来,对其进行鉴定。2003年,贵州省林业专家认为,上甲古榕树群为全国罕见。

相传古时,青年男女在立秋时节来树下祷告祝福,祈求坚贞的爱情。后来这个习俗流传下来,立秋成为了男女青年择偶的佳节,古榕树也成为了他们心目中的媒神。现在,不仅在立秋时节,平常日子,也会有热恋中的男女青年来到这株古树下,祈求上苍给予他们一段美好的恋情。

如今,关岭自治县林业局、岗乌镇政府已经将其列入本地的古树名木名录,并对其进行挂牌保护。

The Most Beautiful Ficus Virens: "Ancient Banyan Woodlot in Shangjia of Guizhou"

Shangjia Buyi Ancient Village Tourist Attraction, Guanling Autonomous County, Anshun City, Guizhou Province, is famous for its "Ancient Banyan Woodlot." There are many ancient Banyan trees in the area, composed mainly of Ficus virens trees, including several that are 1,000 years old. The Ficus virens tree is the most peculiar, featuring a thick stem, unique shape, and hanging roots. Its dense and wide branches extend horizontally to the amazement of visitors.

In recent years, forestry institutions all over the country and even overseas have flocked there to study it. In 2003, Guizhou forestry experts concluded that this Banyan tree group was rare in China.

In ancient times, young men and women came to pray for blessings and faithful love on the Beginning of Autumn (usually falls on August 7 to 9). Later, this became a good day for young men and women to choose their spouse, and the ancient Banyan tree became a special deity in their mind. Now, young men and women in love will come to this ancient tree and pray for harmonious love on any day they like.

The Forestry Bureau of the Guanling Autonomous County and the government of Gangwu Town have introduced special protection for these ancient trees.

最美观光木：大丘"观光木树王"

该树位于福建省建瓯市小桥镇大丘村大夫岭山山脚下。古树伟岸挺拔，傲然屹立于大丘村旁后山，像一名坚强的士兵守卫着村庄的安宁。相传古树为大丘村翁姓始祖所栽。当年他举家迁居经过此地，看重大丘这个地方山高林密、水土丰沃，便扎根于此，待房宇建好后便在屋后种下树木，也就是这株观光木树王。树王经历了800余年的风风雨雨，见证了翁氏子孙在大丘村繁衍生息，留传至今。

（图片摄影 庄晨辉）

The Most Beautiful *Tsoongiodendron Odorum*: "King of *Tsoongiodendron Odorum*" in Daqiu

The tree is located at the foot of Dafuling Mountain in Daqiu Village, Xiaoqiao Town of Jian'ou City in Fujian Province. It stands tall and upright in the back mountain beside Daqiu village like a strong soldier guarding it. It is said that some 800 years ago, when a man named Weng moved his family to where Daqiu Village is today, he found dense forests, fertile soil and rich water resources in the mountain. He decided to plant a tree behind his house, producing what we see as the King of *Tsoongiodendron Odorum* today.

(Photos by Zhuang Chenhui)

中文名：观光木

拉丁名：*Tsoongiodendron odorum Michelia odora* Chun

所在地：福建省建瓯市小桥镇大丘村

树龄：800 年

胸（地）围：503 厘米

树高：3230 厘米

冠幅（平均）：2510 厘米

Chinese Name: 观光木

Latin Name: *Tsoongiodendron odorum Michelia odora* Chun

Location: Daqiu Village, Xiaoqiao Town, Jian'ou City, Fujian Province

Tree Age: 800 years

Chest (Floor) Circumference: 503 cm

Tree Height: 3,230 cm

Crown Width (Average): 2,510 cm

最美南紫薇：莲花山南紫薇

该树位于福建省将乐县万全乡陇源村际头自然村莲花山上。

将乐县有"紫薇之都"的雅誉。境内有3处天然南紫薇古树群落，其中面积最大的一处在白莲镇铜岭村，面积最小一处在万全乡陇源村际头自然村莲花山，只有约50亩，这株南紫薇就生长在这里。近观这棵南紫薇树，只见古树参天，枝繁叶茂，可谓是将乐紫薇之都的"名门望族"。据林业专家考证，这棵树至少已生长了近1600年，为世间罕有。

这棵南紫薇王的附近，曾发生过战争，山顶的古炮台遗址便是见证。南紫薇王的旁边，还有古代留下来的用来保护它的围砌痕迹。

（图片摄影　黄海）

The Most Beautiful *Lagerstroemia Indica*: *Lagerstroemia indica* in Lotus Hill

The tree is located on the Lotus Hill, Jitou Natural Village, Longyuan Village, Wanquan Township in Jiangle County, Fujian Province. Jiangle County is known as "the capital of crape myrtle." Inside the county are three ancient *lagerstroemia indica* woodlots. The largest is in Tongling Village of Bailian Town, and the smallest, covering an area of 3.3 hectares, is located in Jitou Natural Village, Longyuan Village, Wanquan Township. This *lagerstroemia indica* is located in the latter. A close look at the ancient tree reveals that it has luxuriant branches and leaves. According to research of forestry experts, this tree has been growing for about 1,600 years, which is rare in the world.

(Photos by Huang Hai)

中文名：南紫薇
拉丁名：*Lagerstroemia subcostata* Koehne
所在地：福建省将乐县万全乡陇源村
树龄：1580年
胸（地）围：766.2厘米
树高：2900厘米
冠幅（平均）：2130厘米

Chinese Name: 南紫薇
Latin Name: *Lagerstroemia subcostata* Koehne
Location: Longyuan Village, Wanquan Township, Jiangle County, Fujian Province
Tree Age: 1,580 years
Chest (Floor) Circumference: 766.2 cm
Tree Height: 2,900 cm
Crown Width (Average): 2,130 cm

最美紫薇：印江紫薇

这株紫薇位于贵州省铜仁市印江土家族苗族自治县永义乡（现更名为紫薇镇）去往梵净山的必经之路上，距离印江县城30千米，距离梵净山自然保护区西线山门5千米。

古树的花红白相间，朵大色艳，十分美丽，但它不结籽，不繁衍。1998年，入选贵州省古、大、珍、稀树名录。因属第三纪孑遗植物，科学界视其为活化石。紫薇一般为小灌木，多栽培于庭院，像这样高大参天的紫薇，目前在国内仅此一株，世界上也只在日本福冈发现比此稍小的另一株。

千年紫薇的奇特之处，可总结为四字：稀、奇、神、灵。稀：第三纪孑遗植物，绝无仅有；奇：树一年开花三次，脱皮一次；神：种子落地不生，枝条嫁接不活，树干、树叶均可入药；灵：当地百姓和过往游客常祭拜该树，神树也荫庇众生。

The Most Beautiful Crape Myrtle: Yinjiang Crape Myrtle

This crape myrtle tree stands on the road linking Yongyi Township (now Ziwei Town), Yinjiang County in Tongren City of Guizhou Province to Fanjing Mountain, being 30 km away from the Yinjiang county seat and 5 km away from the western route gateway to the Fanjing Mountain Nature Reserve.

The flowers of this ancient tree are red and white, big and bright, but do not seed and reproduce. In 1998, it was recorded in the list of ancient, big, and rare trees in Guizhou Province. Because it belongs to the Tertiary relict plants, the scientific community regards it as a living fossil. Crape myrtle is generally a small shrub, mostly cultivated in courtyards. At present, there is only one large crape myrtle in China, and another slightly smaller one in Fukuoka, Japan.

The crape myrtle is known for its being rare, strange, divine and spiritual – It is rare relic plants of the Tertiary period and a strange tree blooming three times a year and peeling off once; it is considered divine with seeds that don't grow on the ground, branches that cannot be grafted, and the trunk and leaves that can be used as medicine; it is spiritual, because local people and tourists often worship it, believing it will bless all beings.

中文名：紫薇
拉丁名：*Lagerstroemia excelsa* (Dode) Chun
所在地：贵州省铜仁市印江土家族苗族自治县永义乡永义村
树龄：1300年以上
胸（地）围：534厘米
树高：3300厘米
冠幅（平均）：240厘米

Chinese Name: 紫薇
Latin Name: *Lagerstroemia excelsa* (Dodde) Chun
Location: Yongyi Village, Yongyi Township, Yinjiang Tujia-Miao Autonomous County, Tongren City, Guizhou Province
Tree Age: over 1,300 years
Chest (Floor) Circumference: 534 cm
Tree Height: 3,300 cm
Crown Width (Average): 240 cm

最美闽楠：永安"百年神树"

在福建省三明市永安市洪田镇政府驻地东南面 3 千米处的生卿村，有一棵已生长 850 年的楠木，当地人称它为"百年神树"。此树高大挺拔，枝繁叶茂，郁郁葱葱，独木成林。

相传几百年前，一位赫赫有名的风水先生经过此地，发现当地民风淳朴，可百姓却非常贫困，便摆卦一算，才知此处阳气过重、亢阳不化造成阴气衰死而钱财不聚。于是，他在村中空旷处栽下一棵楠木，名为"风水树"，待其枝繁叶茂，枝叶把"阳"光遮掩，招来"阴气"，阴阳调和，便是村民发家致富之时。此后，村民生活果然日益富余，而这棵关系着全村风水命脉的"风水树"，也一直被当地人像"神"一样世世代代守护着……

（图片摄影　庄晨辉）

The Most Beautiful Phoebe Bournei: Yong'an's "100-Year-Old Divine Tree"

In Shengqing Village, 3 km southeast of the Hongtian Town government office in Yong'an City, Fujian Province, there is a tall and luxuriant phoebe bournei tree that has been growing for 850 years. The local people call it a "100-Year-Old Divine Tree."

It is said that, hundreds of years ago, a famous Fengshui geomantic master passed through the place and found local people were honest, but lived a very poor life. He made a divination and calculated that an excess and stagnation of Yang (positive principle in nature) led to the decline of Yin (negative principle in nature). Therefore, he planted a phoebe bournei tree in an open land of the village, which he called "Fengshui Tree." When its branches and leaves could cover the Yang, it would attract Yin and harmonize the two, and the villagers would become rich. And in due course, villagers' life did become better, and this Fengshui tree in the village has been protected by the local people as a God for generations.

(Photos by Zhuang Chenhui)

中文名：闽楠
拉丁名：*Phoebe bournei* (Hemsl.) Yang
所在地：福建省永安市洪田镇生卿村
树龄：850 年
胸（地）围：543.2 厘米
树高：3561 厘米
冠幅（平均）：3040 厘米

Chinese Name: 闽楠
Latin Name: *Phoebe bournei* (Hemsl.) Yang
Location: Shengqing Village, Hongtian Town, Yong'an City, Fujian Province
Tree Age: 850 years
Chest (Floor) Circumference: 543.2 cm
Tree Height: 3,561 cm
Crown Width (Average): 3,040 cm

最美桢楠:"桢楠王"

四川省雅安市荥经县云峰寺有两株被誉为"中国桢楠王"的千年古桢楠。

云峰寺寺内由近 200 株参天古桢楠组成的古树群落,是四川乃至全国规模最大的古桢楠园林,其中,尤以天王殿石梯两侧的这两株最壮观。

千年古树源于何时,是谁栽植,没有确切的记载。据当地人世代相传,这个地方原来有一座家庙,供奉的是孔孟,大约在东汉末年成为道观。唐朝时,一位朝圣瓦屋山、峨帽山的僧人,在这里栽下两株杉树和菩提树,并在这里落脚安身,扩建庙宇,道观逐渐变成佛寺,并保留了原来儒道两家的部分特点,后改名太湖石云峰寺。今天,这些高耸入云、枝繁叶茂的古桢楠树木,就是儒、道、佛修行者先后栽植的;两株"桢楠王"是先期儒家人栽植的。

这些古树经历和见证了荥经的历史,也曾险遭不幸。元初,此地遭受兵灾,寺庙被毁,园林被烧。近代,古树群也遭到了明伐暗盗,损失惨重。特别是民国时期,"桢楠王"被当时政府驻军哨兵的烟火点燃,半边树干被烧干。好在没有烧着的半边第二年开始突然猛长,越长越大,越长越直,基本代替了另一半。烧干的部分至今已 60 多年仍未倒下,它是在不断地向人们倾诉惨遭劫难的辛酸史吧。

The Most Beautiful Phoebe Zhennan: "King of Phoebe Zhennan"

In Yunfeng Temple in Yingjing County, Ya'an City of Sichuan Province, there are two ancient phoebe Zhennan earned the title of "King of Phoebe Zhennan."

In the temple, there are some 200 towering phoebe Zhennan trees, forming the largest ancient phoebe Zhennan tree garden. The most spectacular of these ancient trees are two phoebe Zhennan trees flanking the stone steps leading into Tianwang Temple.

No solid records show who planted them. According to what the locals say, there used to be a family temple dedicated to Confucius and Mencius there, which evolved into a Taoist temple in the late Eastern Han Dynasty (25–220). In the Tang Dynasty, a monk on a pilgrimage to Wawu and Emao mountains settled down and planted a fir and a bodhi trees in the area. He then worked hard to expand the temple, transforming it from Taoist to Buddhist worship, but retaining some characteristics of the original Confucianism as well as Taoism period, and later renamed it as Taihu Shiyunfeng Temple. Today, these towering and luxuriant ancient phoebe Zhennan trees planted successively by Confucian, Taoist and Buddhists worshippers and the two King trees still exercised great influence on local customs.

These ancient trees have witnessed the turbulent history of Yingjing. At the beginning of the Yuan Dynasty (1271–1368), the site suffered from a military disaster in which the temple was destroyed and the tree garden was burned. In the period of the Republic of China (1912–1949), the two King Zhennan trees were set alight by the fireworks launched by the sentinels stationed by the government at that time, and half of the tree trunks were burned. Fortunately, the unburned half began to flourish in the second year and later grew taller and straighter, basically replacing the other half. The burned part has not fallen down despite the passage of more than 60 years, telling later generation the bitter history of the disaster.

中文名:桢楠
拉丁名:*Phoebe zhennan* S. Lee et F. N. Wei
所在地:四川省雅安市荥经县云峰寺
树龄:1700 年
胸(地)围:624 厘米
树高:3600 厘米
冠幅(平均):2200 厘米

Chinese Name: 桢楠
Latin Name: *Phoebe zhennan* S. Lee et F. N. Wei
Location: Yunfeng Temple in Yingjing County, Ya'an City, Sichuan Province
Tree Age: over 1,700 years
Chest (Floor) Circumference: 624 cm
Tree Height: 3,600 cm
Crown Width (Average): 2,200 cm

最美玉兰：玉兰花谷玉兰树

这株顶天立地的玉兰树，生长在陕西黑河国家森林公园大蟒河景区玉兰花谷。该树树干高大挺拔，枝繁叶茂，气势雄伟，树形如一把巨伞，26个大枝向四周伸展，绿荫能覆盖约2亩地的范围。每逢阳春三月，绿叶未绽，数万朵紫白色的玉兰花就怒放枝头，满树像白色蝴蝶随风起舞，悠悠的兰香飘满山谷。曾任周至县县尉的白居易在拜过玉兰王之后，留有诗云："紫粉笔含尖火焰，红脂胭染小莲花。芳情香思知多少，恼得山僧悔出家。"该树是中国玉兰中的花魁，已有1200年树龄，堪称世界玉兰王。

The Most Beautiful Yulan Magnolia: Yulan Magnolia in Magnolia Flower Valley

This towering magnolia tree grows in Magnolia Flower Valley of Damanghe Scenic Spot in Heihe National Forest Park of Shaanxi Province. The tree trunk is tall and straight, with 26 luxuriant branches and a majestic crown spreading like a giant umbrella offering shade over an area of 0.7 hectare. Every March, when the leaves begin to turn green, tens of thousands of purple-white magnolia flowers are in full bloom. When viewed from afar, these flowers have the appearance of butterflies dancing in the wind all over the valley. The Tang Dynasty poet Bai Juyi (772–846), who once served as a captain of Zhouzhi County, came to worship the tree king and left a poem profusely lauding the magnificence and charm of the magnolia tree with a history of 1,200 years and revered as the King of World Magnolia.

中文名：玉兰

拉丁名：*Magnolia denudata* Desr.

所在地：陕西黑河国家森林公园大蟒河景区玉兰花谷

树龄：1200 年

胸（地）围：502.4 厘米

树高：2700 厘米

冠幅（平均）：3750 厘米

Chinese Name: 玉兰

Latin Name: *Magnolia denudata* Desr.

Location: Magnolia Flower Valley, Damanghe Scenic Spot, Heihe National Forest Park, Shaanxi Province

Tree Age: over 1,200 years

Chest (Floor) Circumference: 502.4 cm

Tree Height: 2,700 cm

Crown Width (Average): 3,750 cm

最美重阳木：芷江"云树"

重阳木，位于湖南省芷江侗族自治县城南杨溪河与舞水交汇处的岩桥镇小河口村，距县城 3 千米。此树为西汉时期所植，迄今已有 2000 多年的历史，民称"喜树""重阳木"或"千岁树"。因其树干高大，枝叶繁茂，近观浓荫覆地，远望如云参天，又名"云树"。

古树主干中空，上分三枝向外延伸，枝叶交错，蓬松丛杂。每年雨季，各色蝴蝶、白鹭，密集树上，五彩缤纷，云蒸霞蔚，蔚为奇观。

如今，古树的主干内部已空，但树叶仍然茂郁葱茏，夏秋时分，常有村民游客置方桌、凳椅于其内，饮酒赋诗，怡然为乐，有如世外桃源。

据考证，舞水流程 450 千米，重阳木当年在舞水仅植 4 株，并有民谚为证："大哥远住大洪滩，二姐修行榆树湾。三哥充军到奇滩，只有小妹守家园。"如今，其哥姐均已谢世，依旧只有小妹独守家园。

The Most Beautiful *Bischofia Polycarpa*: "Cloud Tree" in Zhijiang

This tree can be found in Xiaohekou Village of Yanqiao Town, at the intersection of the Yangxi River and Wushui waterways in the south of Zhijiang Dong Autonomous County, Hunan Province. Located 3 km from the county seat, the tree was planted more than 2000 years ago in the Western Han Dynasty (202 BCE–8CE). It is known among ordinary folk as the Happiness Tree, Chongyang Tree and Thousand-Year-Old Tree. Because of its tall trunk and luxuriant branches and leaves, it offers dense shade covering large tracts of ground and is therefore called the Cloud Tree.

The trunk of the ancient tree is hollow. It has three branches extending outwards, interlaced with smaller branches and leaves, giving it a fluffy appearance. Every rainy season, butterflies and egrets dance among the dense trees in a colorful, resplendent and magnificent display. In summer and autumn, villagers and tourists often bring tables and chairs to relax and enjoy the tranquil scene.

According to textual research, only four *Bischofia polycarpa* or Chongyang trees were planted in Wushui in ancient times with this only one in existence today.

中文名：重阳木
拉丁名：*Bischofia polycarpa* (Levl.) Airy Shaw
所在地：湖南省芷江侗族自治县岩桥镇小河口村
树龄：2000 年
胸（地）围：1350 厘米
树高：2400 厘米
冠幅（平均）：2850 厘米

Chinese Name: 重阳木
Latin Name: *Bischofia polycarpa* (Levl.) Airy Shaw
Location: Xiaohekou Village, Yanqiao Town, Zhijiang Dong Autonomous County, Hunan Province
Tree Age: 2,000 years
Chest (Floor) Circumference: 1,350 cm
Tree Height: 2,400 cm
Crown Width (Average): 2,850 cm

中国最美古树
The Most Beautiful Ancient Trees in China

最美桂花："九龙桂"

这株 1100 年的桂花树位于福建省浦城县的临江镇水东村杨柳尖，年产鲜花 240 千克。此树由基部分出九枝，枝枝饱满，年年开花，有"九龙桂"之美誉。

相传，嫦娥久居清冷的月宫，思念繁华的人间，于是，在一个桂花盛开的夜晚，她与七仙女各折桂枝飞到人间，刚好看到浦城这片大地，山清水秀人和，田陌相连成片，祥云瑞气环绕，于是喜盈盈地飘然落地，顺手将桂枝插到地上，四处游玩。夜深，众仙子欲拔桂而归，却发现桂枝已生根。仙女归月去，月桂落人间，仙桂从此在人间开花结果。为何又称"九龙桂"呢？据传说，嫦娥当时降临浦城时，手持两枝桂，加上七仙女人手一枝，便成了九龙桂——浦城丹桂的始祖。

（图片摄影 黄庆党）

The Most Beautiful Sweet-Scented Osmanthus: "Nine-Dragon Osmanthus"

This sweet-scented osmanthus tree, planted 1,100 years ago, is located in Yangliujian, Shuidong Village of Linjiang Town in Pucheng County, Fujian Province, with an annual output of 240 kg of flowers. It has nine branches all sprouting from its base part, each growing flowers, hence the nickname "Nine-Dragon Osmanthus."

According to legend, Chang'e, then living in the Cold Moon Palace one day flew down to the Earth together with seven fairies intending to see the land. When they reached Pucheng, heavily scented by sweet-scented osmanthus, they were attracted by its enchanting emerald hills and connected field where peasants were farming harmoniously. They landed and planted the branches of the sweet-scented osmanthus they were holding before setting out to explore. When they returned at night, they found the tree branches they had planted had taken root in the field. The sweet-scented osmanthus thus settled in the earth even when Chang'e and fairies returned to the moon.

Then how come the name of the Nine-Dragon Osmanthus? According to legend, when Chang'e was in Pucheng, she had two laurel branches and each of the seven fairies had one, hence nine laurel branches—Nine-Dragon Osmanthus, which is the ancestor of the sweet-scented osmanthus in Pucheng.

(Photos by Huang Qingdang)

中文名：桂花
拉丁名：*Osmanthus fragrans* (Thunb.) Lour.
所在地：福建省浦城县临江镇水东村
树龄：1100 年
胸（地）围：502.4 厘米
树高：2700 厘米
冠幅（平均）：3750 厘米

Chinese Name: 桂花
Latin Name: *Osmanthus fragrans* (Thunb.) Lour.
Location: Shuidong Village, Linjiang Town, Pucheng County, Fujian Province
Tree Age: 1,100 years
Chest (Floor) Circumference: 502.4 cm
Tree Height: 2,700 cm
Crown Width (Average): 3,750 cm

最美檫木：沙县"仙姑树"

该树生长在福建省沙县高砂镇上坪村大竹自然村后山一片古树群中。古树刚劲挺拔，主干巨大，需四五个成人方能合抱，在福建当属一绝，而该树离地约3米处长有一个硕大的瘤结，又可谓一奇。

据传，新中国成立前，因大竹自然村离城遥远，茫茫林海，道路不通，村里缺医少药，村民生病无法得到医治，备受病痛煎熬。新中国成立后，镇里成立了卫生院，有位何姓女医生经常坚持走路到上坪各自然村为村民检查身体、治疗疾病。由于路途遥远，巡诊的村民点又多，何医生到大竹时经常已是深夜，"山路上深夜打着火把背着药箱赶路的女医生"形象深深地刻在当地村民的脑海里。村民们感其恩情，都认为何医生就是仙姑转世。何医生到大竹时还经常带村民到后山的原始森林里采集草药。这株檫树王的根皮可治疗多种农村的常见病，如祛风除湿、活血散瘀、止血、风湿痹痛、跌打损伤、腰肌劳损、外伤出血等。时间久了，村民遂称檫树王为"何医生树"或"仙姑树"。

（图片摄影 庄晨辉）

The Most Beautiful Sassafras Tzumu: the "Fairy Tree" in Shaxian County

The tree grows amid a group of ancient trees in Dazhu Natural Village in Shangping Village of Gaosha Town, Shaxian County, Fujian Province. The old tree is strong and straight, with a huge trunk which needs four or five adults holding hands to encircle it. The tree has a huge nodule about three meters above the ground.

It is said that, before the founding of the People's Republic of China in 1949, Dazhu Natural Village, which was a community far from the city, situated in vast forests, with no road access. People there suffered from a lack of medicine. In the 1950s, a medical clinic was set up in the town staffed with a female doctor named He who often trudged along various paths to treat the sick. She often had to walk at night carrying a medical kit. As a token of sincere thanks, the villagers called her a reincarnated female fairy. Doctor He often took villagers to pick medicinal herbs in the deep mountain. She told the villagers the barks of this Sassafras tzumu could be used to treat a variety of common diseases, such as relieving metabolic arthritis and removing dampness, promoting blood circulation to disperse blood stasis, hemostasis, rheumatic arthralgia, traumatic injury, lumbar muscle strain, traumatic bleeding, etc. Later, the villagers began to call this ancient tree a "Fairy Tree" or "Doctor He Tree."

(Photo by Zhuang Chenhui)

中文名：檫木
拉丁名：*Sassafras tzumu* (Hemsl.) Hemsl.
所在地：福建省沙县高砂镇上坪村
树龄：1600 年
胸（地）围：743 厘米
树高：3470 厘米
冠幅（平均）：1760 厘米

Chinese Name: 檫木
Latin Name: *Sassafras tzumu* (Hemsl.) Hemsl.
Location: Shangping Village, Gaosha Town, Shaxian County, Fujian Province
Tree Age: 1,600 years
Chest (Floor) Circumference: 743 cm
Tree Height: 3,470 cm
Crown Width (Average): 1,760 cm

最美香榧："中国香榧王"

这株古香榧王生长在浙江省绍兴市诸暨市赵家镇榧王村，高2米左右处分叉为12条粗壮的树枝，最高年产香榧曾达到750千克。

香榧是从榧树自然变异中选出的优良类型或单株经人工嫁接繁殖而成的优良品种，是唯一的栽培树种，会稽山一带是我国香榧的原产地。根据现有资料记载和古香榧树树龄推测，香榧人工栽培应起于南北朝之前，至唐代已有很高的知名度，盛行于宋代，元、明、清时期已有较大规模发展。香榧集食用、药用、油用、材用和观赏于一身，是会稽山区农业文化发展和兴盛的标志符号。嫁接树能成为古树的并不多，香榧也是现存古树中嫁接树最多的，仅诸暨就有4万多株。

诸暨香榧王，2007年入选浙江农业吉尼斯纪录，被称为"中国香榧王"。

The Most Beautiful Chinese Torreya: "King of Chinese Torreya"

This ancient Chinese Torreya grows in Feiwang Village of Zhaojia Town in Zhejiang Province. At a height of two meters, 12 thick branches shoot up from left and right. Its highest annual output of Chinese Torreya reaches 750 kg.

Chinese Torreya is the only cultivated species growing from natural variation or a fine variety of artificial grafting and propagation. The Kuaiji Mountain area is the origin of this tree in the country. According to the extant records, artificial cultivation of Chinese Torreya should have started before the Northern and Southern Dynasties (420–589), and enjoying high popularity in the Tang Dynasty. It was also popular in the Song Dynasty (960–1279), and enjoyed large-scale development in the Yuan, Ming and Qing Dynasties (1271–1911). Chinese Torreya is a symbol of the development and prosperity of agriculture in Kuaiji Mountain area. There are not many grafted trees capable of lasting so long. Chinese Torreya is also numerically the most grafted tree among existing ancient trees, with more than 40,000 in Zhuji alone.

The Zhuji Chinese Torreya King was chosen for inclusion in the *Zhejiang Agricultural Guinness Book of Records* in 2007.

中文名：香榧
拉丁名：*Torreya grandis* Fort. ex Lind. cv. Merrillii.
所在地：浙江省绍兴市诸暨市赵家镇榧王村
树龄：1360年
胸（地）围：926厘米
树高：1800厘米
冠幅（平均）：2600厘米

Chinese Name: 香榧
Latin Name: *Torreya grandis* Fort. ex Lind. cv. Merrillii.
Location: Feiwang Village, Zhaojia Town, Zhuji City, Shaoxing City, Zhejiang Province
Tree Age: 1,360 years
Chest (Floor) Circumference: 926 cm
Tree Height: 1,800 cm
Crown Width (Average): 2,600 cm

■ 最美百日青：临海百日青

这株 880 年的百日青，位于浙江省台州临海市小芝镇中岙村罗树脚下，树干粗壮，要由 4 名成年男子手拉手才能合抱，体态优美，宛如一簇巨大的绿珊瑚，一直被当地村民奉为"镇村之宝"。

据古村宗谱和树下宋代蔡氏古墓碑文刻记记载，古树栽植于北宋末年南宋初年。在中岙村世代村民的呵护之下，长成了现在的参天大树。村民们曾经以为它是罗汉松，后经专家鉴定，才知道它其实是株百日青。

站于树前，整个树身好似几棵树合体盘旋长成，天然就有一股雄浑苍劲的气势。树干上一个个大小不一的树节，显示出岁月的痕迹，更令人称奇的是"胸中有树"，百日青树干里面还长有一棵小松树，树叶却是四季碧绿繁茂，自有一番闲看云天之姿，当属奇观。

The Most Beautiful Bamboo Leaf Pine Overlooking the Sea

This Bamboo Leaf Pine or *Podocarpus neriifolius D. Don* has a history of 880 years. It is situated at the foot of Luoshu in Zhong'ao Village of Xiaozhi Township, Linhai City, Taizhou, Zhejiang Province. Its trunk is so thick that at least four people holding hands are needed to encircle it. Local people regard it as the "Village Treasure" for its likeness to a cluster of graceful huge green corals.

According to the village's genealogy and funerary inscriptions of the Cai Family's tombs in the Song Dynasty (960–1279), this ancient tree was planted in either the late Northern Song Dynasty (960–1127) or the early years of the Southern Song Dynasty (1127–1279). Under the protection of numerous generations of villagers in Zhong'ao Village, it has grown into a towering giant. Previously, villagers mistook it as arhat pine, but later knew from experts that its real name is Bamboo Leaf Pine or *Podocarpus neriifolius D. Don*.

Standing in front of the tree, one may discover that it seems to be a combination of several trees intertwined, exuding a vigorous aura. On the trunks grow burls of different sizes. They are signs of passing times. More amazingly, among the trunks grows a small pine tree, which is green and lush in all seasons. It is really a wonder to appreciate such a fascinating scene.

中文名：百日青
拉丁名：*Podocarpus neriifolius* D. Don
所在地：浙江省台州临海市小芝镇中岙村
树龄：880 年
胸（地）围：610 厘米
树高：1500 厘米
冠幅（平均）：2000 厘米

Chinese Name: 百日青
Latin Name: *Podocarpus neriifolius* D. Don
Location: Zhong'ao Village, Xiaozhi Township, Linhai City, Taizhou City, Zhejiang Province
Tree Age: 880 years
Chest (Floor) Circumference: 610 cm
Tree Height: 1,500 cm
Crown Width (Average): 2,000 cm

最美蚬木：龙州蚬木王

蚬木王，位于广西壮族自治区崇左市龙州县武德乡三联村陇呼屯，广西弄岗国家级自然保护区陇呼片区实验区内，是弄岗保护区的镇山之宝，也是目前已知的广西乃至华南地区最知名、最古老、直径最大的一株蚬木。

蚬木为椴树科蚬木属植物，主要分布在广西西南的石灰岩地区，因其木质坚硬，被称为"铁木"。蚬木王历经2000多年的沧桑，但依然郁郁葱葱地屹立在弄岗保护区的石灰岩山地之上，堪称弄岗地区良好生态环境的活的"纪念碑"。

相传很久以前，蚬木王所在的地方古木森然、荒无人烟。仲夏的一天，老猎人高公带着几个小伙子，追赶一只猎物钻进了这深山老林。他发现此处土地肥沃、水源丰富，便带着几户人家到这里安营扎寨，建立自己的家园。那时，这里山深林密，人烟稀少，野兽异常猖獗，村里的人畜常常遭受猛兽的袭击。每当听到呼救声，高公总是挺身而出，拔刀相助。一个夏日的傍晚，高公打猎刚回到家。忽然，村东头传来声嘶力竭的呼救声，他立即拿起梭镖飞奔而去。只见一只大老虎呲着牙，正要袭击一名中年妇女。高公屏住呼吸，双手握住梭镖，用尽全力向老虎刺去。老虎猛一回头，梭镖不偏不倚地戳进老虎喉中，那位中年妇女得救了，高公也倒在血泊中。乡亲们为高公举行了隆重的葬礼，并将其安葬在村东头的半山腰中。第三天，当人们按本族风俗带着供品前来探坟时，发现高公墓边竟长出了一棵胳膊粗的蚬木。人们惊喜万分，视这棵神奇的蚬木为高公的化身，奉之为"伏虎之仙""镇妖之神"。他们纷纷摆上供品，点香烧烛，稽首叩拜，祈求"神树"保佑全村老少平安。此后的每年三月初三，村里人到这里拜谒高公墓时，都会给这棵"神树"培上一把土，期盼它长成参天大树，以镇住害人的猛兽。说来也怪，自从有了这棵"神树"后，村里再也没有受过猛兽的攻击，百姓平安度日。

随着时间的流逝，"神树"在不知不觉中长大，高公墓则随着"神树"根基的增粗而逐渐消失，只剩了这棵高大挺拔、闻名于世的蚬木王。

中文名：蚬木
拉丁名：*Excentrodendron tonkinense* (A.Chev.) H. T. Chang et R. H. Miao
所在地：广西壮族自治区崇左市龙州县武德乡三联村
树龄：2300 年
胸（地）围：938.8 厘米
树高：4850 厘米
冠幅（平均）：3000 厘米

Chinese Name: 蚬木
Latin Name: *Excentrodendron tonkinense* (A.Chev.) H. T. Chang et R. H. Miao
Location: Sanlian Village, Wude Township, Longzhou County, Chongzuo City, Guangxi Zhuang Autonomous Region
Tree Age: 2,300 years
Chest (Floor) Circumference: 938.8 cm
Tree Height: 4,850 cm
Crown Width (Average): 3,000 cm

The Most Beautiful *Burretiodendron Hsienmu*: A Tree King in Longzhou

The Tree King of *Burretiodendron Hsienmu* is located in Longhu Experimental Areas of Nonggang, National Nature Reserve in Longhutun, Sanlian Village of Wude Township, Longzhou County, Chongzuo City, Guangxi Zhuang Autonomous Region. It is described as the treasure of the experimental area. Up to date, it has been the best well-known, oldest *Burretiodendron Hsienmu* with the longest diameter of its kind in Guangxi and even Southern China.

Pertaining to Chinese linden of the *Burretiodendron hsienmu* family, this type of tree is mainly distributed in the limestone areas of Southwest Guangxi. They are well-known as "Iron Wood" for their hardness. After more than 2,000 years, the Tree King remains luxuriantly green on the limestone mountain in the Nonggang Nature Reserve. It is truly a living "monument" of Nonggang's good ecological environment.

Legend has it that the area around the Tree King *Burretiodendron hsienmu* was an awe-inspiring but deserted place a long time ago. One day in midsummer, an elderly hunter named Gao Gong, together with several young men, chased their prey into the deep forest. They were attracted by the fertile soil and abundant water, so they later set up their homes here and settled down. At that time, the area was so sparsely-populated that wild animals were rampant and often attacked the villagers and their livestock. Whenever Gao Gong heard a voice calling for help, he always rushed over to anyone in distress. One summer evening, he had just returned home after hunting. Suddenly, a scream for help came from the eastern end of the village. He immediately ran into the deep forest with his spear following the sound. Upon arriving, he saw a tiger baring its gums to attack a middle-aged woman. While holding his breath, he held the spear tightly to stab the tiger with all his strength. But the tiger suddenly turned its head, and the spear partially pierced the tiger's throat. The middle-aged woman was saved, but Gao Gong lay in a welter of blood. The villagers held a grand funeral and buried him halfway up the mountain at the eastern end of the village. On the third day after his death, when people came to visit the cemetery with offerings according to convention, they were surprised to see a thick-armed tree growing beside the cemetery. They rejoiced and regarded this magical tree as the incarnation of Gao Gong. They worshiped it as the god suppressing tigers and demons. They lay down offerings, lit incense and candles, and bowed to pray to the "Holy Tree" to bless the whole village. On the third day of every third month of the lunar calendar, when the descendants of the village come to worship at Gao Gong's cemetery, they will place a handful of soil for this "Holy Tree," hoping that it will grow into a giant tree to dispel harmful beasts. What's quite strange is that villagers have led a peaceful life since then, without suffering from any attacks by wild animals again.

As time has passed by, Gao Gong's cemetery has gradually disappeared, but the "Holy Tree" grows stronger. Nothing but the imposing Tree King remains to mark the spot.

最美湖北梣："映泉鸳鸯树"

在大洪山南麓群山怀抱里，湖北省钟祥市客店镇南庄村二组的土地上，星罗棋布生长着对节白蜡树群。其中，500年树龄的有18棵，千年以上树龄的有7棵。其中一棵对节白蜡因两树合生、盘根错节、相互依偎，被当地群众称为"夫妻树"。

话说这一棵两树合生的对节白蜡，是古时郢都（钟祥前称）胜景之一，官称"映泉鸳鸯树"，民称"夫妻树"，在南庄村世代流传着一个传奇故事。

相传1000多年前，南庄村少妇柳舒氏，身背着一个五六岁大的男孩，一手牵着三四岁的女孩，一手提着瓦罐到村口的珍珠泉提水。任凭女孩怎样哭闹，柳舒氏就是不肯放下男孩去背女孩。小女孩的哭声引起了珍珠泉水神的注意，水神化作过路的白胡子老人询问柳舒氏："这姑娘一定不是你亲生的吧？"柳舒氏答道："公公误会了，背上的男孩儿才是夫君与前妻所生，这孩子从小没了娘，我不疼他谁疼呢？"

当晚，柳舒氏梦到白胡子公公对她说，天下即将大旱三年，让她采摘木梓树枝插到田地里，秋天就能长出叫"苕"的果实，教她用苕泥做成苕砖备饥荒。公公反复叮嘱一定不能告诉别人，否则泄漏天机要遭老天的惩罚。

善良的柳舒氏为了让众乡亲共度灾年，冒着泄露天机之险将消息传于村民。因相信柳舒氏人品，村民纷纷效仿她做苕砖备荒。果然，天下大旱，赤野千里，河水断流，泉水枯涸，田地龟裂，饿殍遍野。而南庄村及周边的百姓因柳舒氏的爱心传播，靠苕砖度日躲过一劫。一个夏夜，柳舒氏夫妇和乡亲们去村口纳凉，突然电闪电鸣，狂风大作，柳舒氏夫妇不见踪影。后来，人们发现在柳舒氏夫妇消失的地方并排长出两棵对节白蜡，融为一体，演绎了一场你中有我、我中有你的千古绝恋。南庄村民感恩柳舒氏的牺牲精神，把这两树合一的对节白蜡树视作夫妇两人的化身，称为"夫妻树"，并修寺建庙，世代祭拜。

The Most Beautiful *Fraxinus hupehensis*: "Yingquan Sweethearts Tree"

Lying in the arms of the mountains at the southern foot of the Dahong range, clusters of *fraxinus hupehensis* trees spread all over the territory of Production Group Two in Nanzhuang Village, Kedian Town of Zhongxiang City (Hubei Province), including 18 of at least 500 years old and seven over 1,000 years old. Among them, one *fraxinus hupehensis* tree is reputed as the "Husband and Wife Tree" by local people because it is a combination of two trees intertwined and dependent on each other. The tree looks like a big umbrella providing thick shade over the ground.

It is said that this tree was one of the scenic spots in Yingdu (the ancient name of Zhongxiang) in ancient China. It was officially called the "Yingquan Sweethearts Tree", and the "Husband and Wife Tree" by local people. There is a legend going back generations in Nanzhuang Village about this.

More than 1,000 years ago, a young woman surnamed Liu Shu in Nanzhuang Village came to the Pearl Spring to fill a crock with water at the village entrance. She carried a boy aged five or six on her back and led a girl aged three or four by the hand. No matter how hard the girl cried, the woman refused to put the boy down and carry the girl instead. The girl's crying drew the attention of the Water God of Pearl Spring. He turned into a white-bearded old man who, passing by, asked Liu: "The girl must be adopted, isn't her?" She retorted: "No, you've got it wrong. It's the boy who was born to my husband and his ex-wife. He has been motherless since his childhood. If it wasn't for me, who would take care of him?"

That night, she dreamed that the white-bearded man said to her that the world was about to experience a severe drought for three years, and asked her to pick wood twigs and insert them into the fields. In autumn, she would harvest a fruit called "Chinese trumpet creeper". The old man also taught her how to make food from this to fight against famine. She was repeatedly warned she must not tell others about this secret; otherwise, she would be punished for her disclosure of a secret belonging to heaven.

However, she was so kind-hearted that she told the villagers the secret regardless of the warnings, in order to help them overcome the famine together. Villagers believed in her and made food bricks with Chinese trumpet creepers following her guidance. As predicted, a severe drought really happened, leading to drying up of rivers and springs, cracks in the land and massive deaths. But the people in Nanzhuang Village and surrounding areas survived the drought by relying on the food bricks. One summer night, when villagers were relaxing at the village entrance, bolts of lightning split the sky and a violent wind sprang up. People found that Liushu and her husband were missing. Later, people discovered a pair of *fraxinus hupehensis* trees growing side by side at the place where the couple disappeared. They were intertwined as a reflection of an eternal love between them. With a deep sense of gratitude for the couple's sacrifice, the villagers termed the tree "Husband and Wife Tree" which was the incarnation of the couple. Moreover, they also built a temple to worship them for generations.

中文名：湖北梣
拉丁名：*Fraxinus hupehensis* Chu, Shang et Su
所在地：湖北省钟祥市客店镇南庄村
树龄：1800 年
胸（地）围：1102 厘米
树高：2800 厘米
冠幅（平均）：1400 厘米

Chinese Name: 湖北梣
Latin Name: *Fraxinus hupehensis* Chu, Shang et Su
Location: Nanzhuang Village, Kedian Town, Zhongxiang City, Hubei Province
Tree Age: 1,800 years
Chest (Floor) Circumference: 1,102 cm
Tree Height: 2,800 cm
Crown Width (Average): 1,400 cm

最美新疆野苹果："树龄最长的野生苹果树"

此树位于新源县喀拉布拉镇沃尔托托山上,树龄已有600年,被上海大世界吉尼斯总部认定为"大世界吉尼斯之最——树龄最长的野生苹果树"。该树从地围往上分为5个主干枝,主干枝平均胸径为0.738米。

相传,500年前的沃尔托托山经常有从中原来的商队,赶着马和骆驼驮着物品,经过此处。在一次出行中,商队不幸遇上了风暴,驼队瞬间被风暴卷走,商队中只留下了一名青年和一名当地的哈萨克女子。几天后,粮食和水都已用尽,女子为了救这名中原男子,将出门时带的最后一个"果子"留给了他,自己却遇难了。男子感念女子的深情与大义,留在了沃尔托托山,带着死去的她,向着圣洁的哈班拜峰前行,来到了山的最高处,却终因饥饿和劳累倒下了。最终,他们一起长眠在了美丽的沃尔托托山。

不知过了多少年,沃尔托托山上长出了一棵苹果树,成熟的苹果经常撒落一地,随着山势滚落到各处。渐渐地,周围的山上长满了苹果树,守望着这片神奇的土地。

而那棵最初的苹果树,生命力最为顽强,经过无数风雨,依然矗立在山巅之上。人们在赶路饥渴时,摘果止渴充饥,在树下谈论着古树的传说,谈论着女子和男子的深情大义。经过这里的人们经常在取食(果子)的时候心怀虔诚,以礼待之。渐渐地人们纷纷效仿,路过此处的人们即使不取食物,也会撕下随身的布衣条,系在树上,保佑大树茁壮成长,继续保佑沿途的人们。后来,这种做法演变成了如今哈萨克族传统的树王祈福仪式。此后,人们将此树称为"神树"。

The Most Beautiful Wild Apple Tree in Xinjiang: "The Oldest Wild Apple Tree"

Situated on Voltoto Mountain of Kalabula Town, Xinyuan County, this 600-year-old tree has been recognized by the Shanghai China Records Headquarters as "The Oldest Wild Apple Tree in China". It has five main branches springing from the ground, each with an average diameter of 738 cm.

Legend has it that caravans from the Central Plains often went across Voltoto Mountain 500 years ago, with cargo carried by horses and camels. One time, a caravan was caught in a storm, and camels were blown away in the twinkling of an eye. Only a young man from the Central Plains and a local Kazakh woman survived the storm. A few days later, their food and water were all used up. In order to save the man, the woman gave him the last "fruit" she carried, and she died. With a sense of gratitude for the woman's affection and righteousness, the man stayed on Voltoto Mountain and took her to the holy Habanbai Mountain. Finally, he died from hunger and fatigue at the mountain peak, lying at rest in the beautiful Voltoto Mountain with the woman together.

Many years later, an apple tree grew at the spot. Mature apples were often scattered on the ground, rolling down everywhere along the mountain. Over time, the surrounding mountains had become filled with apple trees, creating a magical land.

However, the original apple tree, which was of the most dogged vitality, still stood on the top of the mountain, despite countless hardships. When passers-by were hungry, they ate the fruits to combat their thirst and hunger under the ancient tree while talking about its legend and the affectionateness and righteousness of the man and the woman. They picked up fruits with reverence and courtesy, which had gradually become an implied convention. Even if passers-by didn't pick up fruits, they would tear off cloth strips and tie them to the tree, praying a blessing for the tree and passers-by. Later, this practice has evolved into which is presently known as the traditional tree blessing ceremony of the Kazakhs. Since then, people call it the "Holy Tree".

中文名：新疆野苹果
拉丁名：*Malus sieversii*（Ledeb.）Roem.
所在地：新疆维吾尔自治区新源县喀拉布拉镇开买阿吾孜村
树龄：600 年
胸（地）围：738 厘米
树高：1180 厘米
冠幅（平均）：1445 厘米

Chinese Name: 新疆野苹果
Latin Name: *Malus sieversii* (Ledeb.) Roem.
Location: Kaimai Awuzi Village, Kalabula Town, Xinyuan County, Xinjiang Uygur Autonomous Region
Tree Age: 600 years
Chest (Floor) Circumference: 738 cm
Tree Height: 1,180 cm
Crown Width (Average): 1,445 cm

中文名：胡桃
拉丁名：*Juglans regia* L.
所在地：西藏自治区日喀则市桑珠孜区年木乡胡达村
树龄：1600 年
胸（地）围：960 厘米
树高：1400 厘米
冠幅（平均）：1700 厘米

Chinese Name: 胡桃
Latin Name: *Juglans regia* L.
Location: Huda Village, Nyangmo Township, Samzhubzê District, Xigaze City, Tibet Autonomous Region
Tree Age: 1,600 years
Chest (Floor) Circumference: 960 cm
Tree Height: 1,400 cm
Crown Width (Average): 1,700 cm

最美胡桃："核桃树王"

西藏自治区日喀则市年木乡胡达村位于国道 318 沿线，距市区 65 千米，村北 3 千米的山坡上，有一株历经千年风雨的大胡桃树，枝叶繁茂，郁郁葱葱，堪称"核桃树王"。相传此树为吐蕃王朝松赞干布祖父赞普达日年斯亲手所植，至今已有 1600 年的历史，仍年产核桃千余斤。除此以外，年木乡境内还有普嘎土林、雅鲁藏布江、黑颈鹤自然保护区等丰富的旅游资源。

年木乡千年核桃树是桑珠孜区少有的树龄 1000 年以上的现存古树，生长在立地条件极为严苛的半山坡，忍受着常年的山风呼啸，历经千年风霜雨雪依然顽强屹立，诠释了不屈的意志和生命的坚强。古树与桑珠孜区著名的寺庙皓寺隔山相对，沾染了寺庙灵气，已然化身百年古刹的守护者，富有浓厚的民族宗教气息，文化底蕴深厚。如今，千年核桃树已成为远近闻名的旅游景点。为了保护好千年古树，当地政府通过建围栏、连通饮水、专人看护等方法多措并举为古树生长创造良好环境，对古树四周进行了绿化，栽植生态防护和经济树种，改善整体环境。

The Most Beautiful Walnut Tree: "King of Walnut Trees"

Huda Village, Nyangmo Township, Xigaze City, Tibet is located along the National Highway No.318, about 65 km from the city. A large walnut tree, with exuberant foliage and vegetation, grows 3 km north of the village. Legend has it that the tree was planted by Zampdari Niansi, the grandfather of Songtsen Gampo (617–650) of the Tubo Kingdom. With a history of 1,600 years, the tree can still produce walnuts of more than 500 kg every year. In addition, Nyangmo Township is also abundant in tourist resources such as Puga Forest, the Yarlung Zangbo River, and the Black-necked Crane Nature Reserve.

The millennial walnut tree in Nyangmo Township is among rare old trees existent for over 1,000 years in Samzhubzê District. It grows on a hillside with extremely harsh site conditions and has been exposed to year-round mountain gusts. The tree still stands tenaciously despite being battered by the weather, symbolizing the unyielding will and the strength of life. Standing opposite the well-known Haosi Temple in Samzhubzê District, this ancient tree has absorbed the spirit of the temple and become its guardian with a strong ethnic and religious character and a profound cultural property. Until now, the millennial tree has become a renowned tourist attraction. In order to protect it well, the local government have built a fence around it, provided accessible water and sent special people to take care of it, creating a favorable environment for the growth of the ancient tree. Moreover, they also take actions to ensure greenery around the tree, planting economic tree species and trees for ecological protection, thus improving the overall environment.

■ 最美三球悬铃木："其娜尔"古树

在新疆维吾尔自治区和田地区墨玉县阿克萨拉依乡古勒巴格村境内，有一株千年"其娜尔"——悬铃木。这株古悬铃木从主干生出的7大干枝伸向四周，个个粗壮挺拔，竞相媲美。

每当仲夏之际，放眼望去，古树如同一个巨大的绿色蘑菇，镶嵌在村落之间。夏季赤日炎炎，而树荫下则凉风拂面。

在当地群众中流行两种传说：一种传说是，该树是800多年前，由一个名叫艾力伯克阿的伊斯兰传教士从麦加带回国栽种的。久而久之，其顽强的生命力，得到了群众的敬仰，视其为圣树加以保护。来到这里，好客的维吾尔村民会和善地告诉你：绕着这树走7圈，你会如同古树一样长寿。

另一种传说是，在大漠有一位威望很高的统治者，收复了大漠的众多部落，但是，部落赫拜尔城却不屈服于他的统治。在劝降无果的情况下，这位统治者决定攻打赫拜尔城。当军队到达赫拜尔城下时，屡次攻城均以失败告终。这时，身为统治者女婿的艾力将军向其岳父请战，并承诺自己单枪匹马可以攻下此城。军中将领不服，统治者也为女婿的安危担忧，就故意为难他说："你能将身边这棵大碗粗的梧桐树连根拔起就允许你去攻城，否则，你就不要再提此事。"艾力将军当即双手抓住树干一用力，面不改色，将树连根拔起，举过头顶，折成两节当空扔了出去。其中一节落到了阿哈塔木村（今喀拉喀什镇）成活。60年后，一村民建房将此树砍掉了，另一节落到古勒巴格村（今阿克萨拉依乡），也活了，即为现在其娜民族风情园的其娜尔——法国梧桐树。据传，当年艾力将军用了7天7夜终于攻下了赫拜尔城。

当时古勒巴格村的村民发现，这棵飞来的"其娜尔"光秃秃的树桩竟然奇迹般地活了，而且每天都从顶部发出一个新枝，长了7天7夜，生成了7个枝后就不再分叉。而且直到千年后的今天，依然是7个丫枝，一枝不多一枝不少，刚好7个人手拉手合抱才能把它围住。

The Most Beautiful *Platanus Orientalis* Linn.: "Chinar" Ancient Tree

A millenarian *Platanus orientalis* linn called "Chinar" grows in Gulebarge Village, Aksaray Township, Moyu County of Hotan Prefecture in the Xinjiang Uyguar Autonomous Region. Its seven branches extending from the trunk stretch out in all directions, all of them sturdy and tall.

In midsummer, you may find the ancient tree appears to be a huge green mushroom inlaid between the villages, bringing a fresh breeze to people seeking its shelter from the sun.

There are two legends popular among local people. It was said that an Islamic missionary named Ali Bakah introduced it from Mecca and planted it in China more than 800 years ago. Over time, its tenacious vitality caused it to be revered by the people, who protected it as a holy tree. Here, the hospitable Uygur villagers will kindly tell travelers: Walking around the ancient tree seven times brings longevity.

According to the other legend, a prestigious ruler in the desert regained power over many neighboring tribes, except for one in Hebar City, which refused to submit itself to his rule. After surrender negotiations proved no avail, the ruler decided to attack the fortified city. However, numerous attacks ended in failure. At this juncture, General Aili, the ruler's son-in-law, volunteered to join the fight and promised that he could conquer the city alone; this however, aroused the displeasure of other generals in the army. Concerned about his safety, the ruler deliberately embarrassed him and said: "You will have approval to attack the city as long as you can uproot the tree as thick as a big bowl. Otherwise, you should not mention it again." Without hesitation, General Aili grabbed the trunk with both hands, and, even without trying, uprooted it and raised it over his head. Then, he broke it into two parts and threw them up in the air. One part fell down on Ahatam Village (present-day Karakash Town) where it continued to grow. Six decades later, a villager cut it down for building a house. The other part in Gulebag Village (present Aksarayi Township) also survived, currently growing into "Chinar" in Qina Ethnic Culture Park - a French Firmiana simplex . It was said that it took seven days and seven nights for General Aili to finally succeed in the conquest.

At that time, residents of Gulebage Village discovered the bare stump had survived miraculously, and new branchs sprouted from the top over a period of seven days and seven nights – one branch in each single day. A thousand years later, this ancient tree still has seven branches flourishing. And it takes seven people hand in hand to encircle the trunk.

中文名：三球悬铃木
拉丁名：*Platanus orientalis* L.
所在地：新疆维吾尔自治区和田地区墨玉县阿克萨拉依乡古勒巴格村
树龄：882 年
胸（地）围：315 厘米
树高：3200 厘米
冠幅（平均）：3250 厘米

Chinese Name: 三球悬铃木
Latin Name: *Platanus orientalis* L.
Location: Gulebarge Village, Aksaray Township, Moyu County, Hotan Prefecture, Xinjiang Uyguar Autonomous Region
Tree Age: 882 years
Chest (Floor) Circumference: 315 cm
Tree Height: 3,200 cm
Crown Width (Average): 3,250 cm

最美梓叶槭：金家冲梓叶槭

湖南省安化县江南镇黄花溪村鹞子尖，是古代安化重要的驿道、商道和茶道，是万里茶道的重要起点。崇山峻岭中众多的古树引人注目，最令人称奇的是金家冲的梓叶槭，已有1500年历史。

这株生长在岩石上的梓叶槭，虽然生长条件艰苦，但千余年来一直枝繁叶茂，气势磅礴，冠似华盖。其根龙盘虎踞，苔藓密布，视岩石为沃土，纳天地之精华，蕴古树之灵气。粗壮的树干需要4名成年人才能将其环抱，变幻奇特的纹理展示的是它历经千年岁月洗礼后的沧桑与秀美。

相传宋嘉定十五年（1222），大臣史弥远把赵与莒（宋理宗）兄弟接到临安，聘名儒郑清之为其老师。同年，赵与莒被立为沂王，改名贵诚，并领邵州防御使潜邸邵陵（今邵阳）。嘉定十七年（1224），宋宁宗弥留之际，史弥远联合杨皇后密谋废太子，立沂王赵贵诚为新帝。宋理宗征诣京师，日夜兼程从邵阳防地赶往临安，正好遇上邵河（资江源头）涨水，于是，宋理宗一行抄近走旱路，快马加鞭经新化圳上过龙珠山（原为龙驹山，因宋理宗累死一匹座骑葬此而得名），至安化洞市黄花溪，选附近隐蔽之地小憩。随行士兵卸下盔甲挂在梓叶槭树上，后世称此地为金家冲（原为金甲冲）。在宋理宗死后不久，他的一匹坐骑独自来到黄花溪金家冲，凄凉地站立在古槭树下，长久不愿离去，以至留下了4个深深的马蹄印，至今留存。就在那个时候，梓叶槭树突然间遭受了一次严重的雷击，在离地1米多高的地方被拦腰击断。后另发新芽，长成如今的参天古树。

另传，清乾隆五十六年（1791），邑人陶必铨主修旧治梅城南宝塔，往来乡里筹集资金，来到黄花溪村时，正好遇上鹞子尖茶亭复修。陶必铨欣然写下了《鹞子尖茶引》："《禹贡》荆州之域，三邦底贡厥名，李安溪以为名茶类，窃意吾楚所辖。如今之通山、君山及吾邑，实属产茶之乡……"，并交给当地头人刻碑于鹞子尖主峰。以后官至两江总督的陶澍三次经过古槭树旁，并为鹞子尖茶亭作诗及捐田捐钱。

中文名：梓叶槭
拉丁名：*Acer catalpifolium* Rehd.
所在地：湖南省安化县江南镇黄花溪村
树龄：1500 年
胸（地）围：510 厘米
树高：3350 厘米
冠幅（平均）：3500 厘米

The Most Beautiful *Acer Catalpifolium* Rehd: *Acer Catalpifolium* Rehd in Jinjiachong

Yaozijian in Huanghuaxi Village, Jiangnan Town of Anhua County, Hunan Province, was an important route for couriers, business people and tea dealers in ancient Anhua, and also an important starting point of a long tea route. Among the impressive ancient trees in the mountain range, the most amazing one is the *Acer catalpifolium* Rehd. in Jinjiachong, boasting a history of 1,500 years.

Densely covered with moss, its roots imbibe the essence of heaven and earth, forming the aura of a truly ancient tree. Its trunks are so thick that it takes four men holding hands to encircle it. Its changing and strange textures are a symbol of its beauty weathering thousands of years.

According to legend, in year 1222 (the 15th year of the reign of Emperor Jiading (1208–1224) of the Song Dynasty (960–1279)), Shi Miyuan, the then prime minister, took Zhao Yuju [Emperor Li (1205–1264)] with his brother Zhao Yurui to Lin'an and engaged Zheng Qingzhi, a famous scholar, as his teacher. In the same year, Zhao Yuju was crowned as King Yi and renamed Guicheng, while appointed the defense of Shaozhou to secretly arrive at Shaoling (present-day Shaoyang). In 1224, Zhao Guicheng was enthroned as Emperor Li, as Emperor Ning was dying while Shi Miyuan, in collusion with Empress Yang, annulled the prince. One day, the new emperor was on his way to the capital city Lin'an. It happened that the Shaohe (source of the Zijiang River) was rising. As a result, he had to take a different route. Reaching Longzhu Mountain, the force rested and the soldiers took off and hung their armor on an *Acer Catalpifolium* Rehd tree. Later generations called it Jinjiachong. Shortly after Emperor Li's death, one of his mounts came to Jinjiachong alone, forlornly lingering under the ancient tree, leaving four deep horseshoe marks. At that time, the tree suddenly suffered a severe lightning strike and was broken down in its body at more than one meter above the ground. Later, as new buds sprout, over time it grew into a towering tree.

According to another legend, in year 1791 (the 56th year of the reign of Emperor Qianlong (1736–1796) of the Qing Dynasty), Tao Biquan, a native of the city, took responsibility for the renovation of the south pagoda in Mei Town. He arrived at Huanghuaxi Village to raise funds, when the Yaozijian Tea Pavilion was under restoration. He gladly wrote an article entitled *Yaozijian Tea Introduction*: "Within the territory of Jingzhou, as mentioned in *Yugong* (a geographical work composed in the pre-Qin Time), one of the tributes to the Imperial Court is the tea, which Li Anxi considers good. As far as I am concerned, this is quite our Chu area could produce. Now Tongshan, Junshan and my town have truly become the producing area of tea…", and handed it over to local leaders to carve a tablet on the main peak of Yaozijian. Thereafter, Tao Shu, Governor of Liangjiang (fully referred to in Chinese as the Governor-General of the Two Yangtze Provinces and Surrounding Areas Overseeing Military Affairs, Provisions and Funds, Manager of Waterways, Director of Civil Affairs), passed by the ancient tree three times and wrote poems and donated land and money for the pavilion.

Chinese Name: 梓叶槭

Latin Name: *Acer catalpifolium* Rehd.

Location: Huanghuaxi Village, Jiangnan Town, Anhua County, Hunan Province

Tree Age: 1,500 years

Chest (Floor) Circumference: 510 cm

Tree Height: 3,350 cm

Crown Width (Average): 3,500 cm

最美黄连木：陇南黄连木

甘肃省陇南市武都区五库乡安家坝村有一株黄连木。它曾被暴风撕裂成两半儿，树干犹如巨人的双臂，苍劲有力，呈倒八字形，伸向天空，拥抱蓝天。现在古树内膛已腐朽碳化，膛内空虚，可站4人，仅靠树皮支撑，仍枝繁叶茂，生机盎然。阳春三月，黄连木古树新叶形红，格外耀眼，与大山争彩与桃李争艳，庞大身躯笑傲江湖。

The Most Beautiful *Pistacia Chinensis* Bunge: *Pistacia Chinensis* Bunge in Longnan

There is a *Pistacia chinensis* Bunge in Anjiaba Village of Wuku Township, Wudu District in Longnan City, Gansu Province. The tree was once torn into halves in a storm, and its trunks, which were in the shape of an inverted horoscope, stretched into the sky like the arms of a giant full of vigor and power. At present, the ancient tree has been emptied inside due to carbonization, which allows four people to stand in it. Supported only by barks, the tree is still luxuriantly green and full of vitality. In March, the ancient tree is especially dazzling with new red leaves that are no less fascinating than the mountains, peaches or plums.

中文名：黄连木
拉丁名：*Pistacia chinensis* Bunge
所在地：甘肃省陇南市武都区五库乡安家坝村
树龄：2800年
胸（地）围：920厘米
树高：2300厘米
冠幅（平均）：2000厘米

Chinese Name: 黄连木
Latin Name: *Pistacia chinensis* Bunge
Location: Anjiaba Village, Wuku Township, Wudu District, Longnan City, Gansu Province
Tree Age: 2,800 years
Chest (Floor) Circumference: 920 cm
Tree Height: 2,300 cm
Crown Width (Average): 2,000 cm

中文名：亮叶水青冈
拉丁名：*Fagus lucida* Rehd.et Wils.
所在地：湖南省桑植县八大公山国家级自然保护区
树龄：1500 年
胸（地）围：376 厘米
树高：1800 厘米
冠幅（平均）：2825 厘米

Chinese Name: 亮叶水青冈
Latin Name: *Fagus lucida* Rehd.et Wils.
Location: Badagong Mountain National Nature Reserve, Sangzhi County, Hunan Province
Tree Age: 1,500 years
Chest (Floor) Circumference: 376 cm
Tree Height: 1,800 cm
Crown Width (Average): 2,825 cm

■ 最美亮叶水青冈："千手观音"

此树生长于湖南省桑植县八大公山国家级自然保护区斗篷山林区海拔 1918 米的斗篷山之顶。古树树干粗壮挺拔，在树干 3 米高处开始分叉。分枝枝繁叶茂，呈虬折状向外伸展，冠幅平均 28 米，形似千手观音矗立在斗篷山之巅，环视茫茫林海，保佑众生平安。摄影家、文人骚客、专家学者皆誉称此树为"千手观音"。神奇的传说赋予了亮叶水青冈神秘的力量，带着一身灵气作为千手观音的化身而远近闻名。慕名前来的人们拜谒"千手观音"，祈求学业有成、平安健康、幸福永伴。

■ The Most Beautiful *Fagus Lucida*: "Thousand-hand Kwanyin"

This tree grows on top of the Doupeng Mountain, 1,918 meters above sea level, in the forest area of Doupeng in the Badagong Mountain National Nature Reserve, in (Sangzhi County in) Hunan Province. Its trunks are sturdy and straight, and at a height of three meters above the ground, branches start to diverge. The branches flourish and stretch in every direction creating a crown width of 28 meters on average, resembling a statue of the Thousand-hand Kwanyin standing on top of the Doupeng Mountain, from where it overlooks a vast forest, bringing a blessing of safety to all beings. In the eyes of photographers, poets, experts and scholars, it indeed looks like this Buddhist statue. Fantastic legends have imbued the *Fagus lucida* with great mysterious power. People are attracted to worship and pray for their academic success, peace and health, as well as long-lasting happiness.

最美流苏树：连云港"糯米花树"

在江苏省连云港市海州区朐阳街道朐阳村（孔望山景区龙洞庵），有一株雄性流苏树。

流苏树，花开乳白，形似被水浸泡过的糯米，所以当地百姓叫它"糯米花树"，花制成茶饮，便是"糯米茶"。

这株流苏树坐落在龙洞东侧的龙洞庵庙宇院内大门东侧，与西侧1200年圆柏遥相呼应。据龙洞庵住持说，这株流苏树古树种植于南宋时期，距今已有830年历史。龙洞庵建于北齐武平二年（571），迄今已有1430年的历史，是连云港市最早的名刹之一。宋朝以前，龙洞庵称龙兴寺、龙王庙。孔望山面朝大海，背负平川的锦屏山，千百年来孕育了无数奇花异草、珍稀古木。随着时间的流转，这些日益茁壮的古树也成为了孔望山景色的一部分，赋予了孔望山独特的意境与内涵。

4月中旬，流苏花开，由叶心鼓出花苞，花瓣细长，颜色乳白。微风吹过，淡香袭来，花瓣徐徐而下，就像撒落的粒粒糯米。远观古树，似山中祥云，近看则如皑皑白雪，蔚为壮观。流苏树每年花期10天左右，人们争相前来观赏。

目前，这株流苏树生长状况良好，被确定为国家一级古树名木。

中文名：流苏树
拉丁名：*Chionanthus retusus* Lindl.et Paxt
所在地：江苏省连云港市海州区朐阳街道朐阳村
树龄：830年
胸（地）围：150厘米
树高：960厘米
冠幅（平均）：1150厘米

Chinese Name: 流苏树
Latin Name: *Chionanthus retusus* Lindl.et Paxt
Location: Quyang Village, Quyang Street, Haizhou District, Lianyungang City, Jiangsu Province
Tree Age: 830 years
Chest (Floor) Circumference: 150 cm
Tree Height: 960 cm
Crown Width (Average): 1,150 cm

The Most Beautiful Chinese Fringe Tree: "Glutinous Rice Popcorn Tree" in Lianyungang

There is a male Chinese fringe tree in the Longdong Temple, at the Kongwang Mountain Scenic Spot, in Quyang Village, Quyang Street, Haizhou District of Lianyungang City, Jiangsu Province.

Its milky white flowers resemble water-soaked glutinous rice grains, so local people call it a "Glutinous Rice Popcorn Tree". Tea made from its flowers is known as "Teucrium Manghuaense".

The tree is located on the eastern side of the gate inside the Longdong Temple to the east of Longdong Cave, matching a 1,200-year-old China Savin tree in the west. According to the abbot, this ancient tree was planted during the Southern Song Dynasty (1127–1279), with a history of around 830 years. The Longdong Temple, built in 571, is the earliest in Lianyungang City. Before the Song Dynasty, it was also called Longxing Temple or Dragon King Temple. Facing the sea, Kongwang Mountain has been the home of countless exotic flowers and plants as well as rare ancient trees. As time goes by, these increasingly vigorous trees have also become part of the scenery, creating a unique atmosphere in the mountain.

In mid-April, the Chinese fringe tree blooms, with buds bulging from the center of the leaf. Its slender petals are milky white. A gentle breeze blowing, fragrance spreads all over, and the falling pedals resemble glutinous rice. Looking from afar, you may find the tree like auspicious cloud hidden in the mountains; when you approach it, it looks like white snow, making a spectacular scenery. The blooms last about 10 days every year while people rush to view it.

At present, it is in good condition and has been ranked among national first-class ancient trees.

最美荔枝：漳州"桂枝"

此树位于福建省漳州市的台商投资区角美镇福井村培厝社，虽历经 800 年的沧桑，仍枝繁叶茂，每年结果四五千斤。其果实核小味甜，味道特别，有桂花的香味，因此又名"桂枝"。每年 7 月果实成熟时，村里的小孩儿总是眼巴巴地看着荔枝王。培厝的村民的童年记忆，充满着这种独特的桂花香荔枝味。

相传几百年前，培厝比较偏僻，村里的人病了，都不懂得如何去医治，也没钱去看医、买药。一天，赤脚大仙路过此地，告诉他们村里有棵宝，就是这棵荔枝树，并给村民们传授了用荔枝熬药治病的方法。村民们照办后，果然药到病除，病好如初。从此，村民们便将这棵荔枝树当成宝，在那里修建了休息亭，早晨和傍晚都有老人在那里纳凉、听戏、聊天、泡茶，孩童嬉戏玩耍，充满欢声笑语。

（图片摄影　庄晨辉）

The Most Beautiful *Litchi* Tree: "Sweet-scented Osmanthus *Litchi* Tree" in Zhangzhou

This ancient tree is located in Peicuo Community, Fujing Village, (Jiaomei Town) Taiwanese Investment Zone, in Zhangzhou City, Fujian Province. The 800-year-old tree still thrives with a yield of 2,000–2,500 kg of fruits every year. The sweet fruit features a small kernel and a special taste of sweet-scented osmanthus, so it is also known as the "Sweet-scented Osmanthus *Litchi* Tree". When the fruits are ripe in July, children in the village always look wistfully up at the "*Litchi* King". It is the unique fragrance of *Litchi chinensis* mixed with the taste of sweet-scented osmanthus that is always among childhood memories of residents of Peicuo Village.

According to legend, for a long time the village was relatively remote. Villagers didn't know how to get medical treatment when they were sick, nor did they have money for doctors or medicines. One day, the Barefoot Immortal (a figure in Chinese folklore) passed by, telling them that the *litchi* tree was a treasure. He also taught them how to cure diseases with *litchi*. Following his guidance, villagers managed to cure the sick themselves. Ever since, they regard the tree as a treasure; villagers even built a rest pavilion there, where every morning and evening, elders enjoy the cool, listen to traditional operas, chat and make tea. Children play around there, filling the air with their laughter.

(Photos by Zhuang Chenhui)

中文名：荔枝
拉丁名：*Litchi chinensis* Sonn.
所在地：福建省漳州市台商投资区角美镇福井村培厝社
树龄：800 年
胸（地）围：720 厘米
树高：1901 厘米
冠幅（平均）：2470 厘米

Chinese Name: 荔枝
Latin Name: *Litchi chinensis* Sonn.
Location: Peicuo Community, Fujing Village, Jiaomei Town, Taiwanese Investment Zone, Zhangzhou City, Fujian Province
Tree Age: 800 years
Chest (Floor) Circumference: 720 cm
Tree Height: 1,901 cm
Crown Width (Average): 2,470 cm

最美木棉：中山堂木棉

此树位于广东省广州市越秀区中山纪念堂。木棉是广州市的市花，在花城往事中扮演着重要角色。而广州市最古老的木棉王就坐落在具有深厚历史文化底蕴的中山纪念堂的东北角。古树树高约23米，胸（地）围5.9米，虽然年代久远，但是依然雄伟如初。每当春天来临，中山纪念堂的木棉王万花绽放，姹紫嫣红。

348年高龄的木棉王见证了广州山河田海的沧桑巨变，见证了中山纪念堂发生的重大历史事件，如抗战时期广东地区日军在中山纪念堂签字投降，国内外领导人及海内外贵宾来访中山纪念堂、北京奥运火炬传递、亚运会会徽发布及各种纪念孙中山先生活动等。

木棉王脚下有一块立石，上面刻着广州原市长朱光的题词："广州好，人道木棉雄。落叶开花飞火凤，参天擎日舞丹龙。三月正春风。"这首词正是中山纪念堂木棉王的最佳写照。

The Most Beautiful Kapok Tree: Kapok Tree in the Sun Yat-sen Memorial Hall

This ancient tree is located in the grounds of the Sun Yat-sen Memorial Hall in Yuexiu District, Guangzhou City, Guangdong Province. Kapok, the city flower of Guangzhou, plays a leading role in the city history. Guangzhou's oldest kapok tree is situated at the northeast corner of the Sun Yat-sen Memorial Hall, one of the profound historical and cultural heritages of Guangzhou. The ancient tree is about 23 meters high, with a chest (floor) circumferenc of 5.9 meters. Old as the kapok tree is, it is still as majestic as ever. When Spring arrives, the kapok tree is in full bloom, forming dazzling colors.

The 348-year-old kapok tree is a witness of the vicissitudes that Guangzhou has experienced in its expansion, and also of major historical events revolving around the Sun Yat-sen Memorial Hall, such as the signature and surrender ceremony of the Japanese Army in Guangdong at the end of China's War of Resistance Against Japanese Aggression, visits of domestic and foreign leaders and distinguished guests, relay of Beijing Olympic Torch, and release of the emblem of the Asian Games, as well as a wide range of activities honoring Sun Yat-sen.

At the feet of the tree stands a stone with an inscription written by Zhu Guang, the former mayor of Guangzhou, which reads: "Guangzhou is a good place to appreciate kapok. The kapok in full bloom seems like a flying phoenix, towering into the sky and dancing like a dragon amidst the spring breeze in March." This is truly the best portrayal of the old kapok tree in the Sun Yat-sen Memorial Hall.

中文名：木棉
拉丁名：*Bombax malabaricum* DC.
所在地：广东省广州市越秀区中山纪念堂
树龄：348 年
胸（地）围：596.6 厘米
树高：2300 厘米
冠幅（平均）：3485 厘米

Chinese Name: 木棉
Latin Name: *Bombax malabaricum* DC.
Location: Sun Yat-sen Memorial Hall, Yuexiu District, Guangzhou City, Guangdong Province
Tree Age: 348 years
Chest (Floor) Circumference: 596.6 cm
Tree Height: 2,300 cm
Crown Width (Average): 3,485 cm

最美古梅："潮塘宫粉"

广东省梅州市梅县区城东镇潮塘村的"千年古梅",树龄1010年,是广东至今发现树龄最老的梅花,属真梅系直枝梅类宫粉型花梅,为梅花专一品种,被全国古梅专家王其超先生定名为"潮塘宫粉",并已载入《梅国际登录年报2000》。有专家认为这株古梅是宋梅,极具保存、研究、观赏价值。

潮塘山区土地贫瘠,缺水少肥,千年古梅仅其一树傲然挺立。据测算,古梅树高约11米,冠幅12.5米,主干直径103厘米,距地面50厘米处分成双干,两干直径分别为59厘米和44厘米。梅树花色粉红、重瓣、馨香,花径约2.5厘米,花期为每年12月中旬至次年1月下旬,长达50天之久。

据潮塘村民说,最好的赏梅时间为每年冬季大寒前后,这时古梅花开绚烂,婀娜多姿,遍枝粉嫩嫣红、灿若霞锦,醉人的淡红,典雅极致的"宫粉",令人如痴如醉,流连忘返。村民经常在树下观赏、游乐。"潮塘宫粉"恣意怒放,盛况空前,吸引着江西、福建,广东深圳、广州等众多游人纷纷前来观赏。

中文名:梅
拉丁名:*Armeniaca mume* Sieb.
所在地:广东省梅州市梅县区城东镇潮塘村
树龄:1010年
胸(地)围:325厘米
树高:1100厘米
冠幅(平均):1250厘米

Chinese Name: 梅
Latin Name: *Armeniaca mume* Sieb.
Location: Chaotang Village, Chengdong Town, Meixian District, Meizhou City, Guangdong Province
Tree Age: 1,010 years
Chest (Floor) Circumference: 325 cm
Tree Height: 1,100 cm
Crown Width (Average): 1,250 cm

■ The Most Beautiful Ancient Plum: "Powder in Chaotang"

The "Millennial Plum", boasting a history of 1,010 years, is located in Chaotang Village of Chengdong Town, in Meixian District of Meizhou City, Guangdong Province, and is the oldest of its kind found so far in Guangdong. Named as "Powder in Chaotang" by Mr. Wang Qichao, an expert in ancient plum trees, it has been included in the *Annual Report of Plums Registered Internationally* (2000). Some experts believe that the plum can be dated back to the Song Dynasty, which makes it of great value in preservation, research and appreciation.

Chaotang mountainous area ia characterized by barren land and shortage of water and fertilization. Of countless plums over a millennium, only this one manages to weather wind and snow and stand proudly till now. It is about 11 meters high, with a crown width of 12.5 meters and a main trunk whose diameter is of 103 cm . The trunk diverges into two from 50 cm above the ground, with diameters of 59 cm and 44 cm, respectively. The *mume* blossoms, double-petaled, are pink and fragrant, with a flower diameter of about 2.5 cm. The blooms last for 50 days, from mid-December to late January of the next year.

According to Chaotang villagers, the best time to appreciate *mume* blossoms falls around Dahan ("Great Cold", the 24th solar term) each winter. At that time, the *mume* is in full bloom, as brilliant and colorful as rosy clouds. Nothing is more graceful and fascinating as the extremely elegant blooms of the *mume*. It creates a truly intoxicating landscape to linger on. Villagers often admire sceneries and play under the tree. Its beauty has been an attraction for many tourists from Jiangxi, Fujian, Shenzhen, Guangzhou, and so on.

最美枳椇：南雄拐枣

这株树龄为 500 年的枳椇树位于广东省南雄市坪田镇迳洞村，它高大挺拔，雄伟奇峻，树势优美，枝叶繁茂，叶大荫浓，果梗虬曲，状甚奇特，是一级古树。

枳椇又名拐枣，早在《诗经·小雅》中就有"南山有枸"的诗句；《辞源》中解释："枸即枳椇，南山谓之秦岭。"这些历史记载，均体现出枳椇在我国栽培利用的历史久远。

苏联一位学者对拐枣做过不少研究，他认为拐枣在地球上的历史已超过 500 万年，是地球上最古老的果树之一。

The Most Beautiful Raisin Tree: Nanxiong Honey Raisin Tree

This 500-year-old raisin tree is located in Jindong Village of Pingtian Town in Nanxiong City, Guangdong Province. Majestic, straight and graceful, the tree forms a dense shade with its luxuriant branches and tangled stems. It has been ranked among national first-class ancient trees.

Honey raisin tree is evidenced by a verse: "On the southern hill is *ju*" in *The Book of Songs: Xiao Ya*; *Chinese Etymology Dictionary* explains: "*Ju* is officially known as raisin tree, while the southern hill is Qinling." The historical record has demonstrated the long history of the cultivation and utilization of rasin trees in China.

A scholar in the former Soviet Union did much research into honey raisin tree. He believes it to be one of the oldest fruit trees, with a history of more than five million years.

中文名：枳椇
拉丁名：*Hovenia acerba* Lindl.
所在地：广东省南雄市坪田镇迳洞村
树龄：500 年
胸（地）围：472 厘米
树高：2820 厘米
冠幅（平均）：2570 厘米

Chinese Name: 枳椇
Latin Name: *Hovenia acerba* Lindl.
Location: Jingdong Village, Pingtian Town, Nanxiong City, Guangdong Province
Tree Age: 500 years
Chest (Floor) Circumference: 472 cm
Tree Height: 2,820 cm
Crown Width (Average): 2,570 cm

最美米椎:"岭南第一大椎"

此树位于广东省韶关市始兴县深渡水瑶族乡坪田村,当地百姓称之为"米椎王",至今仍能结果,是当之无愧的树王。古树长势旺盛,树形奇特,板根超大,具有很高的美学价值。且该树附近还有10多株胸径50厘米以上的古树。"米椎王"最大的特点是其板状根延伸出地面几米,并高出地面1米多,是至今为止岭南地区发现的最古老的米椎树,华南农业大学教授称其为"岭南第一大椎"。

The Most Beautiful Mizhui Tree: "The Largest Mizhui Tree in Lingnan"

Reputed as the "King of Mizhui Trees" by local people, it is situated in Pingtian Village of Shendushui Yao Ethnic Township of Shixing County, Shaoguan City, Guangdong Province. This ancient tree can still bear fruits, making it worthy of the title. It grows vigorously, with a peculiar shape and super-large roots, bearing high aesthetic value. Most uniquely, its plate-shaped roots extend out for several meters, and grow over one meter above the ground. So far, it is the oldest Mizhui tree found in Lingnan area, and is titled the "Greatest Mizhui Tree in Lingnan" by professors from South China Agricultural University.

中文名:米椎
拉丁名:*Castanopsis carlesii*(Hemsl.)Hay
所在地:广东省韶关市始兴县深渡水瑶族乡坪田村
树龄:1000年
胸(地)围:880厘米
树高:3000厘米
冠幅(平均):3900厘米

Chinese Name: 米椎
Latin Name: *Castanopsis carlesii* (Hemsl.) Hay
Location: Pingtian Village, Shendushui Yao Ethnic Township, Shixing County, Shaoguan City, Guangdong Province
Tree Age: 1,000 years
Chest (Floor) Circumference: 880 cm
Tree Height: 3,000 cm
Crown Width (Average): 3,900 cm

最美人面子："人面子王"

广东省四会市罗源镇石寨村，是一个拥有500多年历史的古堡村寨。石寨村一直有栽植人面子的习俗，有"人面子之乡"美誉。这株"老寿星"人面子树，树高25米，胸围6.7米，冠幅32米，树干要4人才能合抱。此树由江氏先祖江晦岩于公元1470年由四会高街尾江巷迁徙来石寨村时栽植，至今枝叶繁茂，苍劲挺拔，年年开花结果，年产人面子果500千克。

该树虽已有546年历史，但是生长旺盛，树体高，树冠大，树叶密，树干非常壮硕，5个分枝很健康，板根美观，独木成林，因此也被称为"人面子王"。

The Most Beautiful *Dracontomelon Duperreanum* Pierre: "King of *Dracontomelon Duperreanum* Pierre Trees"

Shizhai Village in Luoyuan Town of Sihui City, Guangdong Province, is an ancient castle village with a history of more than 500 years. Local people have followed the practice of planting *dracontomelon* trees, claiming the village is the "Home of *Dracontomelon Duperreanum* Pierre Trees". This ancient tree is 25 meters high, with a maximum circumference of 6.7 meters and a crown width of 32 meters. Its trunk is so thick that at least four people linking hands is needed to encircle it. It was planted by Jiang Huiyan, an ancestor of the Jiang family, when he migrated from Weijiang Alley, Gaojie, Sihui to Shizhai Village in 1470. The tree is still luxuriantly green and full of vigor, with an annual yield of fruits of 250 kg.

Up to date, it still grows vigorously despite 546 years of vicissitudes, its body high, crown big, leaves dense, trunk strong, and with five branches being graceful and luxuriantly green, the tree itself makes a scenery of an entire forest, earning its title as the "King of *Dracontomelon Duperreanum* Pierre Trees".

中文名：人面子
拉丁名：*Dracontomelon duperreamum* Pierre
所在地：广东省四会市罗源镇石寨村
树龄：546 年
胸（地）围：670 厘米
树高：2500 厘米
冠幅（平均）：3200 厘米

Chinese Name: 人面子
Latin Name: *Dracontomelon duperreanum* Pierre
Location: Shizhai Village, Luoyuan Town, Sihui City, Guangdong Province
Tree Age: 546 years
Chest (Floor) Circumference: 670 cm
Tree Height: 2,500 cm
Crown Width (Average): 3,200 cm

最美板栗：怀柔古板栗

　　这株树龄超过 700 年的板栗树，位于北京市怀柔区九渡河镇西水峪村，胸围 5.18 米，平均冠幅 14 米，树高 9 米。

　　怀柔素有"中国板栗之乡"的美誉，板栗栽培历史悠久。清代《日下旧闻考》中记载，"栗子以怀柔产者为佳"。司马迁曾在《史记》中对幽燕地区盛产栗子有过记述，唐代怀柔板栗被定为贡品，辽时曾设立"南京板栗司"管理板栗生产。在明代中期，朝廷投入大量人力物力广植树木而"构筑"了另一道"绿色长城"。皇帝敕命，于边外广植榆柳杂树以延塞马突袭之迅速，内边则开果园栗林以济饥寒之戍卒。怀柔古板栗正是栗林中的一棵，粗大的树干裂为三瓣却又落地支撑后分为三株长成，树干中可同时站上四五人。

The Most Beautiful Chinese Chestnut Tree: Ancient Chestnut Tree in Huairou

This chestnut tree, with a history of more than 700 years, is located in Xishuiyu Village of Jiuduhe Town in Huairou District, Beijing. It is nine meters tall, with a girth of 5.18 meters, and crown width averaging 14 meters.

Huairou, a district of Beijing, boasts a time-honored history of cultivation of Chinese chestnuts, proving itself worthy of its reputation as the "Home of Chinese Chestnuts". As recorded in the book entitled *Research on Old Stories in Beijing* (a book composed in the Qing Dynasty): "The chestnuts from Huairou are the best." According to *Records of the Grand Historian* written by Sima Qian (c.145 or 135–86 BC), Youyan Area was abundant in chestnuts. In the Tang Dynasty, the chestnuts produced in Huairou were designated as tributes to the imperial court. In the Liao Dynasty, the "Nanjing Chestnut Division" was established for the management of chestnut production. During mid-Ming Dynasty, the imperial court invested a lot of manpower and resources to plant trees, building a defense line known as another "Green Great Wall". The emperor ordered elm and willow trees to be planted widely along the borders to impede horse raids by wild minority tribes, and plant orchards of chestnuts to relieve the hunger of garrison soldiers. This Chinese chestnut tree in Huairou District is a remnant of the former chestnut forest. Its thick trunk is split into three parts, which all continue to grow luxuriantly after holding firmly in the earth. At least four or five people can stand on the trunks at the same time.

Chinese Name: 板栗
Latin Name: *Castanea mollissima* BL.
Location: Xishuiyu Village, Jiuduhe Town, Huairou District, Beijing
Tree Age: 700 years
Chest (Floor) Circumference: 518.1 cm
Tree Height: 900 cm
Crown Width (Average): 1,400 cm

中文名：板栗
拉丁名：*Castanea mollissima* BL.
所在地：北京市怀柔区九波渡河镇西水峪村
树龄：700 年
胸（地）围：518.1 厘米
树高：900 厘米
冠幅（平均）：1400 厘米

中国最美古树
The Most Beautiful Ancient Trees in China

最美酸枣：北京"酸枣王"

此树位于北京市东城区花市枣苑住宅小区（原北京市崇文区上堂胡同14号院）内。

此树原生长于崇文区上堂胡同14号的四合院里，因花市街道改建为住宅小区，故被原地保护。2001年10月，在酸枣树附近，挖出一口古井，井内有50个辽金时代百姓常用的打水器皿鸡腿瓶和锥形瓶，表明这里曾是金中都东郊的一个居民点。据《北京志·市政卷·园林绿化志》记载，此树树龄800年，被誉为"酸枣王"，列为北京市一级保护古树，人称"活化石"。"酸枣王"存活至今，世所罕见。酸枣树从金代一路走来，经过几百年的风风雨雨，遭遇雷击、风霜侵蚀而不死，历明清两代几次冻灾而幸存，依然枝繁叶茂、春花秋实，尤为珍贵，人皆以为吉祥树。

The Most Beautiful Wild Jujube: "King of Wild Jujube" in Beijing

This tree is located in Zaoyuan Residence Community, Huashi, Dongcheng District, Beijing, but was originally cradled in the Courtyard No. 14, Shangtang Hutong, Chongwen District, Beijing. It is protected in situ as a result of reconstruction of Huashi Street into residential communities. In October 2001, an ancient well was found near the jujube tree. A total of 50 water containers commonly used by the people in the Liao and Jin Dynasties, such as chicken drumstick bottles and conical flasks were unearthed, indicating that this area was a residential community in the eastern suburbs of Beijing, the then "Central Capital" of the imperial court of the Jin Dynasty. According to the records of *Beijing Chronicles · Municipal Volume · Records of Landscape and Greening*, this tree, with a history of 800 years, enjoys a laudatory title of the "King of Jujube Trees", ranking among the first-class protected ancient trees of Beijing as a "living fossil". It has survived so far, which is rarely seen in the world. Despite hundreds of years of vicissitudes, the tree has survived lightning strikes, wind and frost erosion, and still remains luxuriant through several freezing disasters in the Ming and Qing Dynasties. It is truly a miracle, rendering it to be an auspicious tree for people.

中文名：酸枣
拉丁名：*Ziziphus jujuba* Mill.var.*spinosa*（Bunge）Hu ex H.F.Chow
所在地：北京市东城区花市枣苑
树龄：800年
胸（地）围：230厘米
树高：1500厘米
冠幅（平均）：1100厘米

Chinese Name: 酸枣
Latin Name: *Ziziphus jujuba* Mill.var.*spinosa* (Bunge) Hu ex H.F.Chow
Location: Zaoyuan Residence Community, Huashi, Dongcheng District, Beijing
Tree Age: 800 years
Chest (Floor) Circumference: 230 cm
Tree Height: 1,500 cm
Crown Width (Average): 1,100 cm

最美槲树："菜树奶奶"

此树位于北京市怀柔区宝山镇对石村玄云寺院内。

槲树又名波罗叶，是北方地区荒山造林树种，幼叶可饲养柞蚕。此树是北京市唯一的一株古槲树。目前，古槲树在怀柔区园林绿化局和当地政府、村委会的管护下花繁叶茂，树冠饱满，年年硕果累累，当地人称之为"菜树奶奶"。其果实俗称橡实，果内含有淀粉和蛋白质养分，虽不好吃，却可果腹救命。

The Most Beautiful *Quercus Dentata*: "Grandma of Oak Trees"

This tree is located in Xuanyun Temple, in Duishi Village, Baoshan Town, Huairou District, Beijing.

The *quercus dentata*, also known as Baltic Leaves, pertains to a tree species for afforestation in barren mountains in the northern China, and its young leaves can be used to raise tussah. It is the only ancient oak tree extant in Beijing. At present, the tree is growing in full bloom and fruitful under the management and protection of Huairou District Bureau of Forestry and the Parks of Beijing Municipality, as well as local government and village committee. It has been given the title of "Grandma Oak Tree" by local people. Its fruits are commonly known as acorns, which contain nutrients such as starch and protein. Although the fruits are not tasty, they can save lives.

中文名：槲树
拉丁名：*Quercus dentata* Thunb.
所在地：北京市怀柔区宝山镇对石村
树龄：1000 年
胸（地）围：377 厘米
树高：1570 厘米
冠幅（平均）：1900 厘米

Chinese Name: 槲树
Latin Name: *Quercus dentata* Thunb.
Location: Duishi Village, Baoshan Town, Huairou District, Beijing
Tree Age: 1,000 years
Chest (Floor) Circumference: 377 cm
Tree Height: 1,570 cm
Crown Width (Average): 1,900 cm

最美青檀："檀公古树"

此树位于安徽省池州市青阳县酉华镇二酉村老屋组的村南头，2014年7月，被安徽省人民政府命名为"安徽省名木"，属一级保护古树。

"檀公古树"树龄千年，盘根遒劲，冠若祥凤，虽历经风雨洗礼，时代变迁，但依然枝繁叶茂，郁郁葱葱。"檀公古树"朝北伸出的一根大树枝，紧压着树下古塔的塔顶。大树南边有一口井，井水千年不枯，甘甜醇厚。

据史料记载，以前树边有北宋先人修建过的老"檀公庙"依偎在古树旁边，后来古树不断生长，树根把整座小庙全部包容到树根之下。据董氏宗谱记载，"檀公树"自清康熙四十一年（1702）董姓先人自泾县来此居住时就耸立在村头。董氏家族落户以后繁衍生息，人丁兴旺，家族不断壮大，到清乾隆年间，董家村不论是经济还是教育都发展到鼎盛时期。这时候董氏宗族把"檀公树"长势的盛衰，看成关乎董氏家族兴衰存亡的标志。为报答神树庇护，董姓族人捐钱，于清乾隆癸亥年（1743）八月十五日重修了"檀公古塔"，以示对"檀公古树"的敬畏和感恩。当年董村闺女出嫁，媳妇迎娶，官员巡访，棺椁入土，所乘车轿、所抬棺木进出董村都要绕"檀公古树"一周，一表敬意，二讨彩头，三保平安，四泽子孙。每逢黄道吉日，节气更替，婚丧嫁娶，董氏族人都来焚香祭拜，以示诚心。正月舞龙灯的时候，要先来"檀公古树"绕树一圈，然后各家各户才能接龙。在树下"试水点睛"让龙附有灵性，确保一方水土平安、风调雨顺。后来随着战争爆发，"檀公古树"无人呵护，便一蹶不振，日渐萧瑟发黄。

2014年，青阳县林业部门投入10万余元，对"檀公古树"进行了有效保护，改善了其生长环境。

The Most Beautiful Wingceltis: "Ancient Wingceltis Tree"

The tree is south of Laowuzu, Eryou Village of Youhua Township in Qingyang County, Chizhou City, Anhui Province. In July 2014, it was rated as a "Famous Tree of Anhui Province" by the People's Government of Anhui Province, and is put under first-class national protection.

Despite an age of thousands of years, the tree is still luxuriant, sturdy and vigorous, with its crown looking like an auspicious phoenix. A large branch of the tree protrudes northward, pressing against the top of an ancient tower. To the south of the tree is a well full of sweet and mellow water that has served people for thousands of years.

According to historical records, beside the tree stood Tangong Temple built in the Northern Song Dynasty. As time went by, the tree gradually surrounded the temple with its roots. According to the genealogy of the Dong Family, the tree had been there when the family moved to the village from Jing County in the 41st year (1702) of the reign of Qing Emperor Kangxi. The family settled down and expanded. During the reign of Emperor Qianlong, the village reached its heyday in terms of economy and education. Regarding the Tangong Tree as something crucial that determined the family's rise and fall, the Dong Clan rebuilt the ancient Tangong Pagoda to show their awe and gratitude for the long-lived tree on August 15, 1743. At that time, the clan always made a circle around the ancient tree in case of any wedding, funeral or other celebration/commemorative event to pray for peace, good luck while showing respect for the tree and hoping that it can bless the offspring. They also burned incense on auspicious days and to mark the solar terms in similar tributes, in order to show their sincerity. In the first month of the lunar year, the villagers would walk around the tree to endow the "dragon" with spirituality before performing the dragon lantern dance to seek safety and a good harvest. However, the peaceful and prosperous life was ruined by war. The ancient tree was left unattended, withering little by little.

In 2014, the Forestry Bureau of Qingyang County invested more than 100,000 Yuan in improvement of the growth environment for the ancient tree.

中文名：青檀
拉丁名：*Pteroceltis tatarinowii* Maxim.
所在地：安徽省池州市青阳县酉华镇二酉村
树龄：1000 年
胸（地）围：880 厘米
树高：1700 厘米
冠幅（平均）：1900 厘米

Chinese Name: 青檀
Latin Name: *Pteroceltis tatarinowii* Maxim.
Location: Eryou Village, Youhua Township, Qingyang County, Chizhou City, Anhui Province
Tree Age: 1,000 years
Chest (Floor) Circumference: 880 cm
Tree Height: 1,700 cm
Crown Width (Average): 1,900 cm

■ 最美七叶树：安康七叶树

这株七叶树位于陕西省安康市岚皋县溢河乡高桥村著名的南宫山景区，景区海拔700米。树龄508年，树高27米，胸围3.45米，冠幅27米。金秋时节，树叶由绿转红，在阳光的照耀下，绚丽多彩。

■ The Most Beautiful Chinese horse-chestnut (*Aesculus chinensis* Bunge) in Ankang

The tree is located in the famous Nangong Mountain Scenic Spot in Gaoqiao Village, Yihe Township, Langao County, Ankang City, Shaanxi Province, with an elevation of 700 meters. It is 508-year-old and 27 meters high, with a girth of 3.45 meters, and a crown width of 27 meters. In the golden autumn, its leaves turn red, creating a colorful spectacle in the sunshine.

中文名：七叶树
拉丁名：*Aesculus chinensis turbinata* Blume
所在地：陕西省安康市岚皋县溢河乡高桥村
树龄：508 年
胸（地）围：345.4 厘米
树高：2700 厘米
冠幅（平均）：2700 厘米

Chinese Name: 七叶树
Latin Name: *Aesculus chinensis turbinata* Blume
Location: Gaoqiao Village, Yihe Township, Langao County, Ankang City, Shaanxi Province
Tree Age: 508 years
Chest (Floor) Circumference: 345.4 cm
Tree Height: 2,700 cm
Crown Width (Average): 2,700 cm

■ 最美沙梨："梨树王"

此树位于安徽省宿州市砀山县园艺场六分场场部，树高6.5米，干生九大主枝。4月繁花遮天蔽日，8月硕果金珠坠地，年产量达2000千克。

相传清朝乾隆皇帝下江南，有一次途中行宫就设在砀山县良梨镇境内的訾庄寺院，地方官殷勤地献上当地的特产——砀山酥梨。乾隆皇帝品尝了郭楼村产的酥梨后赞不绝口，当即口谕："捎带为皇考贡品。"第二天乾隆游览梨园时，看到这棵高大健壮、姿态非凡的梨树，深以为奇，随命名为"梨树王"。从此，"梨树王"之名不胫而走。如今，郭楼更名为郭庄，訾庄尤在，而寺院却没有了，只留下了一段脍炙人口的故事。

每年四月，梨花盛开，一簇簇，一层层，似云似雪，铺天盖地，满树冰肌玉骨，花白如银，缠裹掩映。清风徐来，花枝招展，缕缕清香，沁人心脾。梨枝虬曲如龙，披白灿灿一身银装，洒清甜甜漫天寒香，好一派乌龙披雪的美景！盛夏七月，果树郁郁葱葱，绿叶满目青翠，成为绿的海洋。金秋九月，累累硕果挂满枝头，成为果的世界。酥梨果实硕大，黄亮美观，黄澄澄的大酥梨发出诱人的香味，轻咬一口，皮薄多汁，酥脆爽口，香浓味甜。冬季，满园古树肌肤苍黑，铁干嶙峋，乌鳞斑驳，枝丫虬劲，横空逸出。

2014年7月，这株沙梨入选安徽省名木，由省人民政府挂牌保护。2018年"梨树王"被评为"中国最美沙梨"。

中文名：沙梨

拉丁名：*Pyrus pyrifolia*（Burm.f.）Nakai

所在地：安徽省宿州市砀山县园艺场六分场

树龄：230年

胸（地）围：318厘米

树高：650厘米

冠幅（平均）：1600厘米

The Most Beautiful Sand Pear Tree: "King of Pear Trees"

The tree is located at the sixth branch farm of the Horticulture Farm in Dangshan County, Suzhou City, Anhui Province. It is 6.5 meters high and has nine main branches. It blooms in April and bears fruit in August, with an annual output of 2,000 km.

According to legend, Qing Emperor Qianlong once temporarily lived in Zizhuang Temple in Liangli Township, Dangshan County during his southern inspection tour. To honor the Emperor, local officials presented him with the local specialty Dangshan Pear from Guolou Village. Emperor Qianlong liked it so much that he gave oral instructions immediately: "Pay tribute from now on." When visiting a pear orchard next day, Emperor Qianlong was attracted by a tall, robust and extraordinary pear tree. He called it the "King of Pear Trees". Since then, the tree has been well-known. Nowadays, Guolou Village is known as Guozhuang Village. Zizhuang still exists while the Temple has gone, and the legend is left.

In the fourth lunar month each year, the pear tree blooms. Clusters and layers of flowers, like clouds and snow, add radiance and beauty to each other. Swinging in the breeze, the flowers give out intoxicating and refreshing fragrance. The branches look like silver dragons dancing in a snowy world. In July, the leaves of the tree make a green ocean. In September, numerous fruits hang from the branches. Fruits look beautiful, and are large, yellow, thin-skinned, juicy, crispy, refreshing, fragrant and sweet. In winter, the tree turns black, dry and rugged, with branches protruding.

In July 2014, the "King of Pear Trees" was rated as a famous tree under protection of People's Government of Anhui Provincial. In 2018, it was reputed as the "most beautiful Sand Pear in China".

Chinese Name: 沙梨

Latin Name: *Pyrus pyrifolia* (Burm.f.) Nakai

Location: Sixth Branch Farm, Horticulture Farm, Dangshan County, Suzhou City, Anhui Province

Tree Age: 230 years

Chest (Floor) Circumference: 318 cm

Tree Height: 650 cm

Crown Width (Average): 1,600 cm

最美文冠果：渭南文冠果

此树位于陕西省渭南市合阳县皇甫庄乡河西坡村。

文冠果是我国特有的一种优良木本食用油料树种，因树顶端叶多为三裂，似文冠，故命名为"文冠果"。历史上人们采集文冠果种子榨油供点佛灯之用。文冠果原产我国北部干旱寒冷地区，在此生长并不多见。如此古老高大的文冠果树，世所罕见，被村民称为"文冠果王"。

The Most Beautiful Shiny-leaved Yellowhorn Tree: Shiny-leaved Yellowhorn Tree in Weinan

The tree is located in Hexipo Village, Huangfuzhuang Township, Heyang County, Weinan City, Shaanxi Province.

Shiny-leaved Yellowhorn is a woody edible oil tree species unique in China. Because its top leaves are usually three-lobed like a crown ("guan" in Chinese), it is also known as "Wenguanguo". The oil extracted from Shiny-leaved Yellowhorn seeds was long used for lighting. Shiny-leaved Yellowhorn is native to the arid and cold regions in northern China, and is seldom seen in Shaanxi. Such a tall and ancient kind is called "the King of Yellowhorn trees" by local villagers.

中文名：文冠果
拉丁名：*Xanthoceras sorbifolium* Bunge
所在地：陕西省渭南市合阳县皇甫庄乡河西坡村
树龄：1700 年
胸（地）围：430 厘米
树高：1000 厘米
冠幅（平均）：6900 厘米

Chinese Name: 文冠果
Latin Name: *Xanthoceras sorbifolium* Bunge
Location: Hexipo Village, Huangfuzhuang Township, Heyang County, Weinan City, Shaanxi Province
Tree Age: 1,700 years
Chest (Floor) Circumference: 430 cm
Tree Height: 1,000 cm
Crown Width (Average): 6,900 cm

中国最美古树
The Most Beautiful Ancient Trees in China

126

最美辽东栎："虎龙古树"

此株辽东栎位于甘肃省定西市岷县蒲麻镇胡虎龙口村龙山山头，被称为"虎龙古树"。该古树为落叶乔木青冈，5月开黄绿色花，花为单性，雌雄同株。其叶子随天气变化而变化，所以还被称为"气象树"。

The Most Beautiful Liaole Oak: "Tiger-Dragon Ancient Tree"

The ancient tree is situated at the top of Longshan Mountain, Huhu Longkou Village, Puma Township, Min County, Dingxi City, Gansu Province. As a deciduous tree, it blooms in May. Its yellow-green flowers are unisexual and monoecious. Its leaves change with the weather, so it is also called a "meteorological tree".

中文名：辽东栎
拉丁名：*Quercus wutaishanica* Mayr
所在地：甘肃省定西市岷县蒲麻镇胡虎龙口村
树龄：668 年
胸（地）围：307 厘米
树高：1100 厘米
冠幅（平均）：600 厘米

Chinese Name: 辽东栎
Latin Name: *Quercus wutaishanica* Mayr
Location: Huhu Longkou Village, Puma Township, Min County, Dingxi City, Gansu Province
Tree Age: 668 years
Chest (Floor) Circumference: 307 cm
Tree Height: 1,100 cm
Crown Width (Average): 600 cm

最美大果榉：邯郸大果榉

此树位于河北省邯郸市磁县陶泉乡北王村炉峰山上，无论是树龄、海拔、胸围、树高、冠幅、生长势、生长环境都十分珍稀，在华北乃至全国十分罕见。大果榉生长在海拔950米平坦处，树干3米高处有4个主枝向四方延伸，生长旺盛，树体硕大，结实量多，覆阴面积300平方米。在其干分枝处有一株自然生长的油松，形成了天然的"榉抱松"景观。大果榉经历无数次自然灾害，仍然枝繁叶茂，健壮地屹立在磁县炉峰山上。

相传公元前206年，刘邦被封为汉中王后，即以汉中为基地，安定巴蜀，收复三秦。3年后，刘邦趁项羽在齐国停留之际，率领诸侯军一举攻占彭城。项羽闻之，急率精兵3万奔袭，击败刘邦联军20余万，刘邦仅率数十骑兵逃脱，逃至炉峰山才得以喘息。后刘邦建立西汉王朝，在炉峰山植树以示纪念。相传东汉光武帝刘秀也曾在此拴马歇脚，树下青石上现留有马蹄印。

The Most Beautiful *Zelkova Sinica*: *Zelkova Sinica* in Handan

The tree is located on the Lufeng Mountain beside Beiwang Village, Taoquan Township, Cixian County, Handan City, Hebei Province. It is a very rare one in North China and even in the whole country in terms of age, elevation, girth, height, crown width, growth potential and growth environment. Growing on a flat place with an elevation of 950 meters, it has four main trunks extending in all directions at a height of three meters. It is vigorous and fruitful, covering a shade area of 300 square meters. There is a naturally growing Chinese Red Pine (Pinus tabuliformis Carr.) at its crotch, so that an amazing scene is naturally formed of the pine embraced by the *Zelkova Sinica*. Through countless rebirths, it is still there, standing robustly.

According to legend, Liu Bang (256–195 BCE) was made "King of Hanzhong" in 206 BCE; since then, with Hanzhong as his base, he pacified "Bashu" and recovered the "Sanqin" region in Shaanxi Province. Three years later, Liu Bang's army captured Pengcheng when Xiang Yu (232–202 BCE) stayed in Qi State. Learning of this, 30,000 elite soldiers under the leadership of Xiang Yu rushed over and defeated more than 200,000 soldiers led by Liu Bang. In the company of only a few dozen cavalries, Liu Bang escaped to Lufeng Mountain. As Liu Bang founded his Western Han regime later, he ordered the planting of trees on Lufeng Mountain to commemorate that special experience. According to another legend, Emperor Guangwu Liu Xiu (5 BCE–57 CE) of the Eastern Han Dynasty once rested there, leaving some horseshoe prints on the bluestone under a tree.

中文名：大果榉
拉丁名：*Zelkova sinica* Schneid.
所在地：河北省邯郸市磁县陶泉乡北王村
树龄：800年
胸（地）围：430厘米
树高：1350厘米
冠幅（平均）：1700厘米

Chinese Name: 大果榉
Latin Name: *Zelkova Sinica* Schneid.
Location: Beiwang Village, Taoquan Township, Ci County, Handan City, Hebei Province
Tree Age: 800 years
Chest (Floor) Circumference: 430 cm
Tree Height: 1,350 cm
Crown Width (Average): 1,700 cm

最美黄栌：泽州"分脉树"

此树生长于山西省泽州县柳树口镇麻峪村外，为山西省目前发现的最大一株黄栌。此树地上部分分为四主干，每个主干需要三四个成人才能抱住，是村里的"分脉树"。人们经常在此烧香祈祷，视其为"神树"。

The Most Beautiful Smoke Tree: 'Divided Tree' in Zezhou

The tree is outside Mayu Village of Liushukou Township in Zezhou County, Shanxi Province. It is the largest Smoke Tree discovered in Shanxi Province so far. It has four main trunks, each of which cannot be fully embraced by less than three or four adults. Thus, it has been given another name "Divided Tree". Local villagers often burn incense and pray in front of the "sacred tree".

中文名：黄栌
拉丁名：*Cotinus coggygria* Scop.
所在地：山西省泽州县柳树口镇麻峪村
树龄：1000 年
胸（地）围：410 厘米
树高：1500 厘米
冠幅（平均）：2100 厘米

Chinese Name: 黄栌
Latin Name: *Cotinus coggygria* Scop.
Location: Mayu Village, Liushukou Township, Zezhou County, Shanxi Province
Tree Age: 1,000 years
Chest (Floor) Circumference: 410 cm
Tree Height: 1,500 cm
Crown Width (Average): 2,100 cm

最美巨柏：西藏柏树王

世界柏树王园林，位于西藏自治区林芝市八一镇东南方向约8千米处的巴吉村，在318国道旁边，是西藏人民政府批准成立的自治区级柏树林保护区。园内有10公顷近1000棵巨柏，平均树高44米，平均直径1.58米，其中最大的一棵被誉为"中国柏科之最"，树高50米，直径5.8米，距今已有3233年的历史，树冠投影面积1亩有余。平均海拔3040米。

巨柏是国家一级保护树种，更是濒危树种，全世界有5种稀有柏树，巨柏就是其中一种。巨柏在藏语里称为"拉薪秀巴"，有"生命柏树""灵魂柏树"之说，被当地群众誉为"神树"。

The Most Beautiful Cypress: King of Cypress in Tibet

The World Cypress King Garden is located in Baji Village, about eight km southeast of Bayi Township in Nyingchi City, Tibet. Beside the National Highway 318, the Garden is an autonomous region-level cypress forest reserve set up upon approval by the Tibet Autonomous Regional People's Government. There are nearly 1,000 giant cypresses in the Garden, covering an area of 10 hectares. The average tree height is 44 meters, with an average diameter of 1.58 meters. Among them, the largest one is reputed to be the King of Cypress in China due to its venerable age (3,233 years). It is 50 meters high, with a diameter of 5.8 meters. Its canopy projection area is more than 0.067 hectare. The average elevation is 3,040 meters.

Giant cypress is an endangered tree species under first-grade State protection, and also one of the five rare cypress species in the world. In the eyes of Tibetans, it is a sacred tree symbolizing life and the soul.

中文名：巨柏
拉丁名：*Cupressus gigantea* Cheng et L.K.Fu
所在地：西藏自治区林芝市巴宜区八一镇巴吉村
树龄：3233年
胸（地）围：1480厘米
树高：5000厘米
冠幅（平均）：2600厘米

Chinese Name: 巨柏
Latin Name: *Cupressus gigantea* Cheng et L.K.Fu
Location: Baji Village, Bayi Town, Bayi Prefecture, Nyingchi City, Tibet Autonomous Region
Tree Age: 3,233 years
Chest (Floor) Circumference: 1.480 cm
Tree Height: 5,000 cm
Crown Width (Average): 2,600 cm

最美沙棘：西藏古沙棘树

错那千年古沙棘林位于西藏自治区山南市错那县曲卓木乡政府所在地——曲卓木村。经过考证，这片天然古沙棘林有800多亩。而曲卓木乡一带，沿娘姆江河谷分布着大约2000多亩的古沙棘林，其中最大的一株沙棘树高约20米，树龄600年以上。

曲卓木乡的沙棘王林，每一棵沙棘树都有着几百年的生长历史。错那的这片古沙棘林，属于野生柳叶沙棘林，当地人称之为"拉辛"，藏语意思为"神魂树"，即魂魄依附的树。据说旧时噶厦政府就曾派出看护员专门照看这片沙棘林。乡里也雇用了一些当地群众来保护这片古林。如今，越来越多的旅游者来到这里游览。

The Most Beautiful Sea Buckthorn: Ancient Sea Buckthorn Tree in Tibet

The Ancient Sea Buckthorn Forest is located in Quzhuomu Village, Quzhuomu Township, Cona County, Shannan Prefecture, Tibet Autonomous Region. Textual research shows the forest covers an area of more than 53.33 hectares. In Quzhuomu Township, there are more than 133.33 hectares of ancient sea buckthorn along the Niangmu River Valley. Among them, the largest sea buckthorn is about 20 meters high and more than 600 years old.

Every sea buckthorn in Quzhuomu Township is several hundred years old. The sea buckthorn trees in Cona County are all wild Willow-Leaved Sea Buckthorn. Local people call it "Racine", which means "soul tree" in Tibetan. It is said that the Gaxag government of old Tibet once designated specialists to take care of the forest. The Quzhuomu Township People's Government also hired some local people to protect this ancient forest. Nowadays, it attracts

中文名：沙棘
拉丁名：*Hippophae rhamnoides* L. inn
所在地：西藏自治区山南市错那县曲卓木乡曲卓木村
树龄：600年以上
胸（地）围：146 厘米
树高：2000 厘米
冠幅（平均）750 厘米

Chinese Name: 沙棘
Latin Name: *Hippophae rhamnoides* L. inn
Location: Quzhuomu Village, Quzhuomu Township, Cona County, Shannan Prefecture, Tibet Autonomous Region
Tree Age: over 600 years
Chest (Floor) Circumference: 146 cm
Tree Height: 2,000 cm
Crown Width (Average): 750 cm

最美桑树：西藏"古桑树王"

此树生长于海拔 2910 米的西藏自治区林芝市林芝镇帮纳村，树龄约 1600 年，树高 7.4 米，胸围 13 米。相传是由吐蕃王松赞干布和文成公主亲手所植，被称为"古桑树王"。其主干体积达 40 立方米，树干基部分 3 枝；2 枝向上，其中一枝地径 1.56 米，一枝地径 1.75 米；另一枝平出，地径 1.01 米，枝长 7.5 米。

桑树树心充实，树叶茂盛，冠幅遮天蔽日，亭亭如盖，气势磅礴。虽年年开花，但无果实，当地藏族群众称它为"布欧色薪"，意为"雄桑树"。此树被村民视为神的化身。

The Most Beautiful Mulberry Tree: King of Ancient Mulberry Trees in Tibet

The tree is situated at Bangna Village of Nyingchi Town in Nyingchi City, Tibet Autonomous Region, at an elevation of 2,910 meters. It is about 1,600 years old and 7.4 meters high, with a girth of 13 meters. According to legend, it was personally planted by the Tubo King Songtsan Gampo (617–650) and Princess Wencheng (625–680). It is reputed as the King of Ancient Mulberry Trees. It has a trunk volume of 40 cubic meters, and three branches at the trunk base. Two branches extend upward, having the ground diameter of 1.56 meters and 1.75 meters respectively. The third branch extends horizontally, with a ground diameter of 1.01 meters and a length of 7.5 meters.

It is full, lush and majestic with a huge crown. It blooms every year, but bears no fruit. Local villagers call it "male mulberry tree", regarding it as an incarnation of a deity.

中文名：桑
拉丁名：*Morus alba* L.
所在地：西藏自治区林芝市林芝镇帮纳村
树龄：1600 年
胸（地）围：1300 厘米
树高：740 厘米
冠幅（平均）：1800 厘米

Chinese Name: 桑
Latin Name: *Morus alba* L.
Location: Bangna Village, Nyingchi Town, Nyingchi City, Tibet Autonomous Region
Tree Age: 1,600 years
Chest (Floor) Circumference: 1,300 cm
Tree Height: 740 cm
Crown Width (Average): 1,800 cm

最美楸树：原平"龙凤古楸"

山西省原平市大林乡柏枝山下的西神头村扶苏庙前，有两株古楸树，人称"龙凤楸"。北边的1株为龙楸，高约30米，胸围13.2米，平均冠幅17.25米；南边的1株为凤楸，高约30米，胸围11米，平均冠幅18米。估测树龄在1500—2000年。龙楸是我国已知楸树中树龄最大、胸径最大的1株，故称"华夏第一楸"。这两株古树虽然历经千年风雨，至今仍枝叶繁茂，古朴苍劲。

"龙凤古楸"干朽枝枯，萌生的新枝组成硕大的树冠，每年5月初，紫花满树，十分壮观。

扶苏庙是当地百姓为了纪念秦始皇之子扶苏神勇忠贞而修建的，始建年代不详，唐贞观年间太宗李世民敕令尉迟敬德督工扩建。唐陶翰《太子崖》中有这样的诗句："扶苏秦太子，举代称其贤。百万犹在握，可争天下权。束身就一剑，千古人共传。"

The Most Beautiful Catalpa Bungei: Ancient Twin Catalpa Trees in Yuanping

There are Twin Catalpa Trees in front of the Fusu Temple in Xishentou Village at the foot of Baizhi Mountain in Dalin Township of Yuanping City, Shanxi Province. The one in the north is reputed to be the Dragon Catalpa, 30 meters high, with a girth of 13.2 meters and an average crown breadth of 17.25 meters. The southern one has the nickname of Phoenix Catalpa, about 30 meters high, with a girth of 11 meters and an average crown breadth of 18 meters. They are estimated to be 1,500–2,000 years old. The Dragon Catalpa is hailed as the Number One Catalpa in China with the largest girth. Despite their venerable age, both trees are still luxuriant, sturdy and vigorous.

The newly sprouted branches form a huge canopy. In early May, the Twin Catalpa Trees are bedecked with purple flowers, looking very spectacular.

The Fusu Temple was built to commemorate the loyal and brave Fusu (?–210 BCE), eldest son of Emperor Qin Shihuang (259–210 BCE). Later, Tang Emperor Taizong (627–649) ordered Yuchi Jingde (585–658) to rebuild it. Tang Dynasty poet Tao Han thought highly of Fusu, exclaiming in a poem that Prince Fusu, the son of Emperor Qin Shihuang, was very virtuous. Although commanding millions of troops, he was so loyal to his father that he killed himself on order from an imperial edict.

中文名：楸
拉丁名：*Catalpa bungei* C.A.Mey.
所在地：山西省原平市大林乡西神头村
树龄：1320 年
胸（地）围：1300 厘米
树高：3000 厘米
冠幅（平均）：1800 厘米

Chinese Name: 楸
Latin Name: *Catalpa bungei* C.A.Mey.
Location: Xishentou Village, Dalin Township, Yuanping City, Shanxi Province
Tree Age: 1,320 years
Chest (Floor) Circumference: 1,300 cm
Tree Height: 3,000 cm
Crown Width (Average): 1,800 cm

最美麻栎：永济橡树

山西省永济市虞乡镇古城西南坡上的张家窑村，有两株大麻栎（又称橡树）。

清乾隆十五年（1750）《文庙崇圣祠重修碑记》载："舜远祖中条之上姚坪山，为其圣土。舜降于山下沩傍之汭流也。弟象受兄事所憾，于古井植之一树，护之于舜，称为象树也。"证明该树系舜弟象所植，并受到舜帝褒扬，遂有"象树"之谓。

近年来，古橡树、沩水、汭水、姚井、舜泉（广孝泉）、虞幕古国、阪之都——虞乡等一系列历史遗存陆续被发掘考证，这些历史遗存对于研究舜降生地，舜迎娶娥皇、女英二妃之地，虞幕古国与虞舜之都的关系，以及舜帝的德孝思想的起源与发展具有重要意义。特别是对于考证舜"耕历山，陶河滨，渔雷泽""嫔于虞""舜都蒲阪"等诸史料的记载，提供了弥足珍贵的实物证据，是今人探究远古历史文化演进以及中国称谓之由来的"活化石"。

The Most Beautiful Sawtooth Oak: Oak Trees in Yongji

There are two sawtooth oaks in Zhangjiayao Village on the southwest slopes of Yuxiang Township in Yongji City of Shanxi Province.

According to the *Inscription on Rebuilding Confucius' Temple* written in 1750 during the reign of Qing Emperor Qianlong, Xiang, the younger brother of Shun (a legendary monarch in ancient China), planted a tree by an ancient well. Later, the tree was named after Xiang-Xiang tree or oak tree in English.

In recent years, many historical relics of ancient oaks have been unearthed in succession at places including Weishui and Ruishui rivers, Yaojing Well, and Shunquan Spring (Guangxiao Spring). They are of great significance to study Shun and his thoughts on virtue and filial piety. They are precious physical evidence and "living fossils" for exploring the history and culture of China.

中文名：麻栎
拉丁名：*Quercus acutissima* Carruth.
所在地：山西省永济市虞乡镇张家窑村
树龄：4200 年
胸（地）围：470 厘米
树高：600 厘米
冠幅（平均）：2000 厘米

Chinese Name: 麻栎
Latin Name: *Quercus acutissima* Carruth.
Location: Zhangjiayao Village, Yuxiang County Township, Yongji City, Shanxi Province.
Tree Age: 4,200 years
Chest (Floor) Circumference: 470 cm
Tree Height: 600 cm
Crown Width (Average) 2,000 cm

最美元宝槭：和顺元宝槭

　　山西省和顺县青城镇神堂峪村，生长着一棵高大茂密的元宝槭，俗称"色树"。此树高 25 米，胸围 5.1 米，主干高 2.1 米，2.1 米以上分为两杈，树干粗壮有力，彰显着它的古老，枝叶三季青翠茂密，暗示着它持久旺盛的生命力。树冠平均冠幅 23 米，遮天蔽日，覆阴面积超过 400 平方米。从一个侧面看，整个树冠恰似一个巨型圆球，村民形象地称它为"绣球"；而从另一个侧面看，树冠又豁然变成两个相切的圆球，头挨头、肩并肩，就像一个连体婴儿。

　　神堂峪村位于和顺县青城镇虎峪沟内，与明朝尚书王云凤（号虎谷）的陵园（虎谷坟）毗邻。在很久以前，人们选村择址的时候，除了认准这里土地肥沃，环境优美外，更主要的是看中了这棵树的存在。这棵树也就顺理成章地成了神堂峪村的风水树、平安树，逢年过节，村民都会来古树前祈福。

The Most Beautiful Acer Truncatum: Acer Truncatu in Heshun County

There is a tall and luxuriant acer truncatum bunge in Shentangyu Village of Qingcheng Township in Heshun County, Shanxi Province. Commonly known as the "color tree", it is 25 meters high, with a girth of 510 cm. Its trunk is 2.1 meters high, thick and robust. It becomes bifurcated at a position 2.1 meters above ground. It is almost evergreen. Its average crown breadth is 23 meters, and its shaded area exceeds 400 square meters. Seen from one side, its crown resembles a giant sphere, and is vividly called "hydrangea" by local villagers. Seen from the other side, the crown looks like two tangent interconnected spheres.

Shentangyu Village is located in Huyugou of Qingcheng Township in Heshun County, and is adjacent to the Hugu Tomb of Minister Wang Yunfeng (1465–1518) of the Ming Dynasty. The site of Shentangyu Village was selected in consideration of its fertile land, beautiful environment and the presence of the most beautiful acer truncatum. The tree is a mascot for villagers, who pray in front of it to celebrate holidays and festivals.

中文名：元宝槭
拉丁名：*Acer truncatum* Bunge
所在地：山西省和顺县青城镇神堂峪村
树龄：500 年
胸（地）围：510 厘米
树高：2500 厘米
冠幅（平均）：2300 厘米

Chinese Name: 元宝槭
Latin Name: *Acer truncatum* Bunge
Location: Shentangyu Village, Qingcheng Town, Heshun County, Shanxi Province
Tree Age: 500 years
Chest (Floor) Circumference: 510 cm
Tree Height: 2,500 cm
Crown Width (Average): 2,300 cm

最美紫丁香："华北最大紫丁香"

此树位于山西省沁水县中村镇下川村村边，枝干向上生长并向四面延伸，枝繁叶茂，形态优美，生长旺盛，犹如乔木，曾被誉为"华北最大紫丁香"。丁香树花期较长，每年6月前后开放，因其冠大花多，缕缕清香随风扑面而来，令人心旷神怡。

令人称奇的是，这株紫丁香不像一般灌木丛生矮小、没有明显的主干，而是主干胸径0.82米、高约5米，枝干呈向上生长的奇特形态，并向四周辐射。站在树下向上看去，丁香树犹如一位老者抬起双臂，为你遮挡炎炎夏日，送来丝丝凉意和缕缕清香。

紫丁香树的树干中空，仅依靠两边树皮生长，后经修复，如今枝繁叶茂，葱葱郁郁。数百年来，紫丁香听松涛，观雪景，赏山花，和鸟鸣，沐浴世间最纯净的阳光雨露，人们看见它，就会产生一种敬畏之情。

The Most Beautiful Lilac: Largest Lilac in North China

This tree stands beside Xiachuan Village of Zhongcun Township in Qinshui County, Shanxi Province. With its branches growing upward and extending in all directions, it is luxuriant, vigorous and beautiful as an arbor. It is hailed as "the Largest Lilac in North China". It blooms around June every year, with a long flowering period. Its large crown and numerous flowers present a fragrant and refreshing scene.

It is not as short as the general shrub. In general, its trunk is short, with a girth of 0.82 meter, and it is about 5 meters high. More amazingly, its branches grow upwards, extending in all directions. Standing underneath, it seems like an old man raising his arms to keep off the scorching sun bringing coolness and fragrance.

Through restoration, its originally hollow trunk has been revitalized, making it luxuriant and vigorous. For hundreds of years, it has harmoniously coexisted with the surroundings, including pines, mountain flowers, and birds. People always feel a sense of awe and veneration when looking at it.

中文名：紫丁香
拉丁名：*Syringa oblata* Lindl.
所在地：山西省沁水县中村镇下川村
树龄：300年
胸（地）围：257厘米
树高：1420厘米
冠幅（平均）：920厘米

Chinese Name: 紫丁香
Latin Name: *Syringa oblata* Lindl.
Location: Xiachuan Village, Zhongcun Town, Qinshui County, Shanxi Province
Tree Age: 300 years
Chest (Floor) Circumference: 257 cm
Tree Height: 1,420 cm
Crown Width (Average): 920 cm

最美银杏：莒县银杏

在山东省最古老的寺院之一日照市莒县浮来山镇定林寺内，有一株参天古树，远看形如山丘，龙盘虎踞，气势磅礴，冠似华盖，繁荫数亩。它就是具有"活化石"之称的"天下银杏第一树"。

据《左传》记载："（鲁）隐公八年（前715）九月辛卯，公及莒人盟于浮来。"会盟距今已有2700多年，专家推断在会盟的时候，这棵银杏树已经是相当大了，差不多有1000多年。由此推断，此树树龄应该超过3700年。

几千年来，这棵古老的银杏树，历经风风雨雨，保持着顽强的生命力。这棵树王的果实和一般的银杏果明显不同，别的树结的银杏果呈纺锤状，大而长，而这棵树的果子又小又圆，味道也特别可口。近几年还出现了一种奇观，就是在树的主干上，也常见生叶结果。另外，在大树的枝丫根部，长出了30个形似钟乳石状的树瘤，轻轻叩击，发出咚咚的声响，里面似乎是空的。

这株古银杏树会发出声响，多发生在春季晴天的晚上，夜阑人静之时，听上去似有人紧闭双唇从鼻腔内发出的声音。其实，发出奇异声响的原因是高大的树干内，部分腐朽了的木质部形成许多孔洞。在光合作用下，树冠抽芽发叶需要大量水分，树干内的无数导管、筛管疏导液体，发出声响，这种声响经腐木孔洞，引起了共鸣。

The Most Beautiful Ginkgo: Juxian Ginkgo

Dinglin Temple in the Fulai Mountain Town of Juxian County is one of the oldest temples in Shandong Province. In its grounds is a towering ancient tree that looks like a hill from a distance. It is so majestic, with a large shaded area formed by a large crown. It is hailed as the "No.1 ginkgo tree in the world" as well as a "living fossil".

According to the records in the *Commentary of Zuo*, Duke Yin of Lu and the envoy from Juxian County met on the Fulai Mountain in the 9th lunar month of 715 BCE, to form an alliance. Some experts infer the ginkgo tree was already almost 1,000 years old before this meeting. In other words, it should be more than 3,700 years old.

Now it still maintains a tenacious vitality today. It is unique in terms of its shape and the taste of its fruit. Ordinary ginkgo fruits are fusiform, big and long. In contrast, this tree's fruit is small and round, but tastes better. In recent years, some leaves and fruits have continued to be seen on its trunk. In addition, there are 30 stalactite-like burls at the roots of branches. When tapped lightly, they give out a rattling sound, suggesting they are empty.

Amazingly, the tree can also make an autonomous sound on spring evenings. It sounds like the voice coming from the nasal cavity. Why? It turns out that a large amount of liquid passes through many holes formed in the decayed trunk to nourish the buds to make the peculiar sound.

中文名：银杏
拉丁名：*Ginkgo biloba* L.
所在地：山东省日照市莒县浮来山镇定林寺
树龄：3700 年
胸（地）围：1582.56 厘米
树高：2500 厘米
冠幅（平均）：3025 厘米

Chinese Name: 银杏
Latin Name: *Ginkgo biloba* L.
Location: Zhendinglin Temple, Fulai Mountain Town, Juxian County, Rizhao City, Shandong Province
Tree Age: 3,700 years
Chest (Floor) Circumference: 1,582.56 cm
Tree Height: 2,500 cm
Crown Width (Average): 3,025 cm

最美红花天料木：霸王岭母生

在海南省昌江县霸王岭国家森林公园内雅加景区天路栈道约 50 米处，有一株 1130 年的红花天料木。历经千年沧桑，它见证了大自然的变迁，也赐予人间苍生平安幸福。海南一些已婚妇女，特别是久婚不孕的妇女会前来祭拜，祈求早生贵子。当地村民及游客上山，会面对神圣庄严肃穆的千年古树，祈求平安，许下心愿，期待得到古树的赐福。

红花天料木别名母生、山红罗、高根、红花母生等。因该树种萌芽力强，母树被砍伐，会从树桩基部萌发出许多幼苗，通常有 3 到 6 条可长成大树，而且越砍伐越长，越长越快。一株红花天料木种下去，可供数代人甚至十几代人砍伐取材，因此有"母生"的别称。为了让其便于萌生枝条，在砍伐时留下较高的伐根，故而也称为"高根"。海南人盖房子主梁木首选母生树，常将其萌生的枝条作为造房用的桁条，以祈求人丁兴旺，子孙发达。

The Most Beautiful Homalium Hainanense: Homalium Hainanense in Bawangling National Forest Park

There is a Homalium hainanense at about 50 meters away from the Tianlu Plank Road in the Yajia Scenic Area of Bawangling National Forest Park of Changjiang County, Hainan Province, estimated to be 1,130 years old. It is regarded as a sacred tree. Local women, particularly those who are infertile, are accustomed to pray before for an early birth. Also, local villagers and tourists pray for peace and blessings, showing their awe and veneration.

Homalium Hainanense has several other names, including Rebirth, High Stump, etc. It has a strong germination capacity. Even if it is cut down, three to six seedlings will grow out of its stump. It has sustained itself through successive cuttings for generations. Therefore, it has earned the name "Rebirth". In order to make it easier to sprout, the higher stump is usually retained, so it is also called the "High Stump". Homalium hainanense is preferred to make the girder, and its branches often used as a focus to pray for a flourishing population.

中文名：红花天料木
拉丁名：*Homalium hainanense* Gagnep.
所在地：海南省昌江县霸王岭国家森林公园
树龄：1130 年
胸（地）围：450 厘米
树高：3800 厘米
冠幅（平均）：1000 厘米

Chinese Name: 红花天料木
Latin Name: *Homalium hainanense* Gagnep
Location: Bawangling National Forest Park, Changjiang County, Hainan Province
Tree Age: 1,130 years
Chest (Floor) Circumference: 450 cm
Tree Height: 3,800 cm
Crown Width (Average): 1,000 cm

最美陆均松:"五指神树"

这株 2600 年的陆均松,坐落在海南省霸王岭国家级自然保护区(白沙县境内)东四林场后山。它的五条分枝,形态酷似一个巨人手掌的五根手指,并且朝同一方向逆时针旋转,故而被称为"五指神树"。

"五指神树"古老苍劲,枝繁叶茂,扭曲的树体凸起一条条鼓足劲的"肌肉",让人肃然起敬,需要 6 个人才能将其合抱。"五指神树"为雄株,相距其 200 米处的山脊另一侧还生长着一株大小相近的陆均松,为雌株。

相传很久以前,当地有一位黎族青年进山打猎,数日不归,其恋人进山呼喊寻找,找寻许多天未见踪影,最终两人化成了两株树,事后发现竟然近在咫尺,让人叹息。

五指神树历经千年沧桑,见证了大自然的变迁,也赐予人间苍生平安幸福,有"霸王归来不看树"的美誉。

The Most Beautiful Dacrydium Pierrei: "Five-Finger Sacred Tree"

This 2,600-year-old tree, Dacrydium Pierrei, stands on the mountain behind the Dongsi Forest Farm in Bawangling National Nature Reserve (part of Baisha County) in Hainan Province. It has five branches like huge five fingers that rotate counterclockwise. Therefore, it is hailed as a "Five-Finger Sacred Tree".

It is luxurious and vigorous, with "muscles" in the twisted trunk. It is very thick, and at least six people linking hands are needed to encircle it. It is staminiferous. On the other side of the ridge 200 meters away, there is a pistillate Dacrydium Pierrei of the similar size.

According to legend, a young man of Li the ethnic group went hunting in the mountains. He had not returned after several days and his spouse anxiously went to search him but was unsuccessful. Finally, the couple turned into two trees close to each other.

Having a long history, they bring people peace and happiness. It can be said that the trees in Bawangling National Nature Reserve outshine all others in the world.

中文名：陆均松
拉丁名：*Dacrydium pierrei* Hickel
所在地：海南省霸王岭国家级自然保护区（白沙县境内）
树龄：2600 年
胸（地）围：720 厘米
树高：2800 厘米
冠幅（平均）：2700 厘米

Chinese Name: 陆均松
Latin Name: *Dacrydium pierrei* Hickel
Location: Bawangling National Nature Reserve, Baisha County, Hainan Province
Tree Age: 2,600 years
Chest (Floor) Circumference: 720 cm
Tree Height: 2,800 cm
Crown Width (Average): 2,700 cm

最美冬青：南阳冬青树

河南省南阳市镇平县城西北的高丘乡刘坟村有一座宝林寺，寺前的半山坡上生长着一棵高大挺拔的小果卫矛。此树苍劲古朴，气势非凡，枝叶青翠欲滴，蕴含生机，当地人称为冬青树。

据推算，此树树龄约有 2000 年。相传东汉时期，光武帝刘秀曾驾宿宝林寺，他的马就拴在这棵冬青树上。此树原在寺院内，被称为"圣树"，备受僧人爱戴。元朝时期，寺院上迁，该树被拒之门外，受到冷落。后来树的右枝端长出一个"虎头"，整棵树看上去如龙似虎，蔚为壮观。于是，古时又传说它是"神仙遗物"，形龙似虎。

据《纲目前记》和《山海经》述，这棵冬青树所处的伏牛山主峰五垛山，古名"骑立山"或"倚帝山"，相传东汉王朝二十八宿多出于此。如今仍有光武帝刘秀的饮马泉、严子陵隐居严陵河畔教书的先师庙，以及五龙二虎逼炎章的古战场"狗家滩"和著名的炎章墓等古代遗迹。

The Most Beautiful Holly Tree: The Holly Tree in Nanyang

Baolin Temple is situated in Liufen Village, Gaoqiu Township, northwest of Zhenping County in Nanyang City of Henan Province. On the hillside in front of the temple stands a tall holly tree, known locally as holy tree. It is luxuriant and vigorous with extraordinary growth momentum.

It is about 2,000 years old. According to legend, Emperor Guangwu (5 BCE–57 CE) of the Eastern Han Dynasty once tied his horse to this holly tree that originally stood inside Baolin Temple, which was moved during the Yuan Dynasty, leaving the sacred tree unattended. Later, a "tiger head" grew out of the right branch, making the tree look very spectacular.

According to *Gang Mu Qian Ji (Introduction to the Compendium of Materia Medica)* and *Shan Hai Jing (The Classic of Mountains and Seas)*, Wuduo Mountain (main peak of the Funiu range) on which the tree stands was formerly known as "Qili Mountain" or "Yidi Mountain". According to legend, most of 28 veteran generals of the Eastern Han Dynasty came from here. Many historical relics have been preserved, including the Yinma Spring (where Emperor Guangwu's horse drank water), Xianshi Temple by the side of Yanling River (where Yan Ziling (39 BCE–41 CE) taught), the ancient battlefield Goujiatan, and the famous Yanzhang Tomb.

中文名：冬青
拉丁名：*Ilex purpurea* Hassk.
所在地：河南省南阳市镇平县高丘乡刘坟村
树龄：2000 年
胸（地）围：292 厘米
树高：1600 厘米
冠幅（平均）：1750 厘米

Chinese Name: 冬青
Latin Name: *Ilex purpurea* Hassk.
Location: Liufen Village, Gaoqiu Township, Zhenping County, Nanyang City, Henan Province
Tree Age: 2,000 years
Chest (Floor) Circumference: 292 cm
Tree Height: 1,600 cm
Crown Width (Average): 1,750 cm

最美皂荚：商丘皂角树

在河南省商丘市虞城县木兰镇小孟楼西头，距离传说中花木兰老家两三千米的地方，长有一株1600年的皂角树。相传花木兰替父从军时，父女就是在这棵树下依依惜别的。

传说当年魏员外喜得长女木兰（花木兰姓魏），十分高兴。木兰长至七八岁时，越发聪明伶俐，读书识字、习武练剑无所不通，被魏员外视为掌上明珠。美中不足的是，木兰从小头发细黄，魏员外夫妇甚是苦恼，四处求医问药，不见效果。一天，魏员外在村中见一老妇满头乌发，上前施礼求教良方，妇人感其爱女心切，赠一树种，嘱其三更种下，来年取其树叶捣碎洗头。魏员外按老妇人指点种下树种，第二年春，小树长成时，遂按照老妇人所说方法给木兰洗头。坚持几次后，木兰头发果然变得乌黑发亮。同村姐妹看到后也都纷纷效仿。后来，小树逐渐长大结出果实，人们就用其果实来洗头洗衣，一直沿袭至今。

The Most Beautiful Honey Locust: Locust Tree in Shangqiu

There is a 1,600-year-old honey locust tree in the west of Xiaomenglou, Mulan Township, Yucheng County, Shangqiu City, Henan Province. It is two to three kilometers away from the hometown of the legendary Hua Mulan. According to legend, Hua Mulan parted reluctantly from her father under the tree.

The story is that the real surname of Hua Mulan was Wei, and her father was a ministerial councilor. At the age of seven or eight, she was extremely skilled in martial arts. However, her hair was thin and yellow since early childhood. Her parents sought medical advice here and there, but gained no favorable result. One day, her father saw an old woman with black hair. Moved by his love of Mulan, she gave him a tree seed and suggested he plant it at midnight. When the sapling grew out next year, he picked some leaves to wash the hair of his beloved daughter. After several washings, her hair turned black and shiny. Other girls in the village followed suit. Later, the sapling grew up and bore fruit. Afterwards, local people used the fruits to wash their hair and clothes and this tradition was passed down.

中文名：皂荚
拉丁名：*Gleditsia sinensis* Lam.
所在地：河南省商丘市虞城县木兰镇小孟楼
树龄：1600 年
胸（地）围：370 厘米
树高：3500 厘米
冠幅（平均）：3200 厘米

Chinese Name: 皂荚
Latin Name: *Gleditsia sinensis* Lam.
Location: Xiaomeng Lou, Mulan Town, Xucheng Village, Shangqiu City, Henan Province.
Tree Age: 1600 years
Chest (Floor) Circumference: 370 cm
Tree Height: 3500 cm
Crown Width (Average): 3,200 cm

最美蒙古栎：千年蒙古栎

在辽宁省沈阳市新民市辽河巨流河段左岸，大喇嘛乡长山子村北山的高地上，有株千年的蒙古栎。据考证，古栎为北宋真宗景德年间所植，历经千年风雨，树干粗壮苍凸，扭曲成各种形神并茂的图案，树干需要3个人才能合抱，树皮拼接着多幅"水墨丹青"画，形神兼备。千年蒙古栎是沈阳地区古树中当之无愧的寿星之王，在周边以杨树、柳树、榆树、槐树等为主体的森林植被环境中，独树一帜。

据专家学者分析，树下的一眼清泉是古树长势不衰的主要原因。据村中老者讲，古树的树洞中有蛇出没，是不能碰的（其实谁也没有看到过蛇，可能是为了不让人破坏古树，才编了这个故事）。据说有个醉汉路经此地被树根绊倒，爬起来勃然大怒，便回家取斧砍树，结果砍了几下后，树干就流了血。醉汉以为是自己出血了，不敢砍了，回到家后就一病不起。至今，树干根部仍有较深斧痕。此后再也没有人贸然砍树了。

古树似有神灵，备受世人崇敬，不少善男信女前来许愿请安，求子求学求财，求功名利禄。一些佛教信徒在此树旁修建的小庙，终年香火不断。

The Most Beautiful Quercus Mongolica: Millennium Mongolian Oak

There is a millennium Mongolian oak on the mountain north of Changshanzi Village in Dalama Township on the left bank of the Liaohe in Xinmin City, Liaoning Province. Textual research shows the tree was planted during the reign of Emperor Zhenzong (968–1022) of the Northern Song Dynasty. It remains luxurious and vigorous. Various patterns featuring unity of form and spirit formed by layers of bark. At least three people are needed to hug the tree with linked hands. It is the oldest tree in Shenyang, looking unique in the surrounding forest vegetation environment in which poplar, willow, elm, and locust trees play a dominant role.

Experts and scholars attribute its unfailing growth to a clear spring underneath. According to village elders, the tree cannot be touched because there are snakes inside it (in fact, this story is made up to prevent people from destroying it). It is said that a drunkard stumbled on a tree root, and became so irritated he cut the tree with an axe. Unexpectedly, the trunk bled after a few chops. Mistakenly thinking he was bleeding, he stopped cutting out of fear. After returning home, he was confined to his bed with a serious illness. So far, axe marks still are visible, and no one has dared to cut it down ever since.

It is regarded locally as a sacred tree. Beside the tree is a temple. Many devout men and women often burn incense in the temple, praying for children, fame and fortune.

中文名：蒙古栎（柞树）
拉丁名：*Quercus mongolica* Fisch. ex Ledeb.
所在地：辽宁省沈阳市新民市大喇嘛乡长山子村
树龄：1000 年
胸（地）围：400 厘米
树高：1100 厘米
冠幅（平均）：1200 厘米

Chinese Name: 蒙古栎（柞树）
Latin Name: *Quercus mongolica* Fisch. ex Ledeb.
Location: Changshanzi Village, Dalama Township, Xinmin City, Shenyang City, Liaoning Province
Tree Age: 1,000 years
Chest (Floor) Circumference: 400 cm
Tree Height: 1,100 cm
Crown Width (Average): 1,200 cm

最美青杨：清原青杨

辽宁省抚顺市清原满族自治县湾甸子镇砍橡沟村有一株古老的青杨，据林业部门测算，这棵树距今已有 500 多年的历史，属杨柳科杨属中的青杨组。通常情况下，由于是速生树种，杨属植物的生命周期一般只有 20 至 30 年，可这棵树不仅从明朝中期一直活到了现在，而且还越发长得茂盛，单单凭着它这么强大的生命力，这棵杨树就足以"世称王"了。

"一株挺拔世称王，耸立浑河古道旁。日照晴岚腾紫气，风摇疏影荡池塘。"这首诗相传是乾隆皇帝回关东祭祖时，曾御笔题诗盛赞这棵亭亭如盖、耸入云天的小叶青杨。作为同类树种当中的"老祖宗"级古木，它正以雍容华贵的姿态挺立在浑河源景区，迎接着每年来此观光的游人。

The Most Beautiful Cathay Poplar: Cathay Poplar in Qingyuan

There is an ancient Cathay poplar in Kanxianggou Village, Wandianzi Township of Qingyuan Manchu Autonomous County in Fushun City, Liaoning Province. It is more than 500 years old. Cathay poplar is a fast-growing tree species, with the average age of only 20 to 30 years. However, this tree has been growing vigorously since the middle of the Ming Dynasty, showing strong vitality. It deserves to be called "the King of Cathay Poplar".

According to legend, Qing Emperor Qianlong returned to the area east of Shanhaiguan to worship his ancestors. He was deeply attracted by a towering Cathay poplar on the bank of the Hunhe, so that he wrote a poem for it. The oldest Cathay poplar stands upright in the Hunheyuan Scenic Area, welcoming numerous tourists in a graceful manner.

中文名：青杨
拉丁名：*Populus cathayana* Rehd.
所在地：辽宁省抚顺市清原满族自治县湾甸子镇砍橡沟村
树龄：500 年
胸（地）围：636 厘米
树高：2410 厘米
冠幅（平均）：2525 厘米

Chinese Name: 青杨
Latin Name: *Populus cathayana* Rehd.
Location: Kanxianggou Village, Wandianzi Township, Qingyuan Manchu Autonomous County, Fushun City, Liaoning Province
Tree Age: 500 years
Chest (Floor) Circumference: 636 cm
Tree Height: 2,410 cm
Crown Width (Average): 2,525 cm

最美赤松：江东赤松王

这株赤松，被当地人称为"神树"，也被称为"启运树"，又称"江东赤松王"或"神树赤松王"，位于辽宁省抚顺市新宾满族自治县木奇镇木奇村。古松气势宏伟，树冠庞大，占地面积为780平方米，树上从无鸟巢，虽经数百年风霜雨雪，但至今枝叶繁茂，四季葱绿，独霸一方。经国家林业部门确认，此树是我国境内年龄最大、最健美的赤松之一，其红枝绿荫相互掩映，自成风景，在苍山秀谷中尤显壮美，远近闻名，是新宾县著名的景点之一。

传说当年努尔哈赤还没有发迹的时候，有一天打猎发现了两只美丽的梅花鹿。这两只鹿边跑边等他，后来把他引到这里就不动了。于是，努尔哈赤搭弓上箭，想把鹿射杀掉。就在他再次寻找目标的时候，鹿消失了，在鹿站立过的地方出现了一棵奇异的松树，通体发着红光。努尔哈赤惊呆了，他缓缓放下弓箭对着这棵松深施一礼，转身走了。从此努尔哈赤就步上了统治中华大地的霸主道路——开创了清王朝。而后的几位政绩显赫的清朝皇帝也先后来这里参拜过。这棵赤松历经时代的演变和清王朝的兴衰，也亲历了近代战争的考验。

抗日战争时期，日军想拿它做炮弹箱子。锯子锯开树的表皮后，涌出一股红色的汁液，日军吓得逃掉了，再也不敢提及伐树之事。

神奇的赤松获得了人们的敬仰。时至今日，附近的人们但凡遇到大事，或命运将有转折的时候，都要先来拜访这棵"神树"，希冀未来更加美好。

The Most Beautiful Red Pine: King of Red Pine in Jiangdong

Local people regard this as a sacred tree. It is also known as the "Fate Tree" and the "King of Red Pine". It is situated at Muqi Village of Muqi Township in Xinbin Manchu Autonomous County, Fushun City, Liaoning Province. Covering an area of 780 square meters, it looks magnificent, with such a huge crown. Not a single bird's nest has ever been found on it. Despite hundreds of years of travails, it is still luxurious and evergreen. It is one of the oldest and most beautiful red pines in China, and a special attraction in Xinbin County.

According to legend, Nurhachi (1559–1626) found two beautiful sika deer when hunting one day. He followed them to a certain place where he prepared to shoot his arrow, only to find that the deer had disappeared. Suddenly, a strange red pine loomed into view. He was so amazed that he bowed to it. He turned away, embarking on the path of establishing the Qing regime. Later, several outstanding emperors of the Qing Dynasty visited the sacred tree, which ultimately saw both the rise and fall of the Qing Dynasty, and also the cruelty of modern wars.

During the War of Resistance against Japanese Aggression, some Japanese soldiers attempted to cut it down to make a cannon box. However, red juice resembling blood flowing out of bark scared them and they ran away.

To local people, it is a holy tree, and they are accustomed to praying for good luck, fame and wealth in front of it.

中文名：赤松
拉丁名：*Pinus densiflora* Sieb.et Zucc.
所在地：辽宁省抚顺市新宾满族自治县县木奇镇木奇村
树龄：1300 年
胸（地）围：370 厘米
树高：2650 厘米
冠幅（平均）：2950 厘米

Chinese Name: 赤松
Latin Name: *Pinus densiflora* Sieb.et Zucc.
Location: Muqi Village, Muqi Town, Xinbin Manchu Autonomous County, Fushun City, Liaoning Province
Tree Age: 1,300 years
Chest (Floor) Circumference: 370 cm
Tree Height: 2,650 cm
Crown Width (Average): 2,950 cm

最美剑阁柏：剑阁翠云廊古柏

在四川省剑阁县汉阳镇翠云廊景区，有株孤独而神秘的大树——2300多岁的剑阁柏，当地百姓称为"松柏长青树"。树皮纵裂、灰褐色，窄长成条片状，不似柏树粗糙，枝叶如松似柏，枝条密，不下垂，远看似松，近看是柏。

民间传说，孙悟空对自己被骗上天庭封为"弼马温"十分不满，因此大闹天宫，在搅乱了王母娘娘的蟠桃宴，酒足饭饱后，思念老巢水帘洞，便扛着金箍棒，一个筋斗云飞上天空，由于酒醉心烦，精神恍惚，不辨路向，匆忙间打翻了天庭玉皇大帝百花园内一株翠柏盆景。翠柏坠落入风景秀丽的大柏树湾，从此就定居凡尘，与世间翠柏和谐而居，茁壮成长，但始终保持着树干是柏，枝叶是松的仙株本色。

据专家考证，剑阁柏在全球仅此一株，被誉为"国之珍宝"。1978年9月，著名树木分类学家赵良能先生认定其是柏木新种，因首次在剑阁发现，定名为"剑阁柏木"。《植物分类学报》在1980年第2期以"柏木属一新种"一文正式定名刊载后，引起了国内外不少林业科学工作者的重视。

中文名：剑阁柏
拉丁名：C.chengiana var jiangensis（N.Chao）
所在地：四川省剑阁县汉阳镇翠云廊景区
树龄：2300 年
胸（地）围：364.2 厘米
树高：2900 厘米
冠幅（平均）：950 厘米

Chinese Name: 剑阁柏
Latin Name: *C.chengiana var jiangensis* (N.Chao)
Location: Cuiyun Corridor Scenic Area, Hanyang Town, Jiange County, Sichuan Province
Tree Age: 2,300 years
Chest (Floor) Circumference: 364.2 cm
Tree Height: 2,900 cm
Crown Width (Average): 950 cm

The Most Beautiful Jiange Cypress: Ancient Cypress in Jiange Cuiyun Corridor Scenic Area

There is a lonely and mysterious big tree in the Cuiyun Corridor Scenic Area, Hanyang Township, Jiange County, Sichuan Province. It is a Chinese Weeping Cypress, more than 2,300 years old. Local people call it the "Evergreen Pine and Cypress". Its gray-brown bark is longitudinally cracked, narrow and flaky. It resembles a pine at a distance and a cypress close up, with dense branches and leaves.

According to legend, the Monkey King created a tremendous uproar in the heavenly palace because of anger at being deceived while working as the Supervisor of Heavenly Horses. After disturbing the peach feast of the Queen Mother of the West, he planned to return to Water Curtain Cave by repeated somersaults. However, due to drunkenness, and distracted by anger, he overturned a cypress bonsai in the flower garden of the Jade Emperor. The cypress fell into the beautiful Cypress Bay. Since then, it has been growing vigorously. It has a trunk of cypress, but the branches and leaves of a pine.

It is unique in the world, and is reputed as a "Treasure of China". In September 1978, famous tree taxonomist Zhao Liangneng identified it as a new species of cypress, and named it as "Jiange Cypress" because it was discovered in Jiange. He published an article in the second issue of the *Journal of Plant Taxonomy* in 1980. It attracted the attention of many forestry scientists at home and abroad.

最美香果树："丁木大仙"

在四川省成都市大邑县森林小镇西岭镇飞水村，有一株五六人才能合抱的千年古树——香果树。该树树形古朴独特，一直被当地人称为"丁木大仙"。

冯家坎世世代代是冯、代两姓聚居地，新中国成立前曾在香果树后建有东岳庙，代冯两姓历代均在庙里祭祀祖先，在香果树下燃香礼拜，祈求风调雨顺，年年平安，由此形成习俗和对"丁木大仙"的礼拜敬奉，故该香果树得以世代留存。

此株香果树庞大的树冠像一把巨伞，山风刮过，巨伞轻轻摇动，在蓝天白云下优哉游哉，好不逍遥自在。粗大的树干上，生长着许多寄生植物，叶片苍翠欲滴，根上长满青苔，显出了大树的古老苍劲。

在盛夏酷暑少花时节，"丁木大仙"却怒放满树繁花，宛如"幽谷香兰"闪烁于青翠的群山之中。虽历经风吹雨打、严寒酷暑，但依然生机勃勃。在人们心中，它不单单是一棵树，更有着几辈人儿时的记忆，它象征着一种精神，一种不屈不挠的生命力。当地村民爱护着它，敬奉着它，守护着它，让它永葆青春，成为永远的乡愁。

The Most Beautiful Emmenopterys Henryi: 'Dingmu Grand Immortal'

The millennium tree is situated at Feishui Village, Xiling Township in Dayi County of Chengdu City, Sichuan Province. At least five or six people are needed to encircle it with their arms. It is simple and unique in shape, and is deemed as a holy tree.

Before liberation, the Dongyue Temple stood behind the tree. The Feng and Dai clans were accustomed to burning incense under the tree, praying for peace and good harvest. It has been well-preserved for generations.

It has a canopy like a huge umbrella that sways in a gentle and leisurely way against the prevailing mountain wind. There are many parasitic plants in its thick trunk. Green leaves, and moss around the roots highlight its history and vigor.

In the midsummer season when there are few flowers, it blooms like "Chinese cymbidium", shining in the verdant mountains. Despite many travails, it is still full of vitality. Symbolizing a determined spirit and tenacious vitality of eternal youth, it mirrors the memories of many generations. Local villagers always cherish and protect it.

中文名：香果树
拉丁名：*Emmenopterys henryi* Oliv.
所在地：四川省成都市大邑县西岭镇飞水村
树龄：1000 年
胸（地）围：763.02 厘米
树高：3580 厘米
冠幅（平均）：2300 厘米

Chinese Name: 香果树
Latin Name: *Emmenopterys henryi* Oliv.
Location: Feishui Village, Xiling Town, Dayi County, Chengdu City, Sichuan Province
Tree Age: 1,000 years
Chest (Floor) Circumference: 763.02 cm
Tree Height: 3,580 cm
Crown Width (Average): 2,300 cm

■ 最美柏木：七贤古柏

在四川省北川羌族自治县永安镇大安村一座舒缓的山坡前，一株巨大古柏树因其 10.3 米的胸围，成为仅次于陕西省黄帝陵古柏的全国第二大古柏树，吸引了不少村民和游人参观祭拜。

巨树周遭 10 多米的范围内，散落伴生着 7 株 2 米粗细的"小"古柏树，交相辉映间不仅形成独木成林的奇特景观，又因主干七根、伴生七株的"双七"树群而被赋予各种传奇代代流传。这便是北川"七贤柏"。

关于七贤柏，民间流传着一个故事：一日，庙住持让几个小和尚在庙旁栽种几株柏树，小和尚们便一株一株地栽植起来，很快种了 7 棵柏树苗。但是有一个小和尚懒惰贪玩，在其他人辛苦种树的时候只顾躲在一边玩耍。过了一会儿天气忽变，大雨将至，这个小和尚便急忙将剩余的 7 棵柏树苗种在了一起。没想到，小和尚的偷懒，反而催生了一株神奇的"七贤柏"。

"七贤柏"曾有几次差点被砍伐的经历，不过都因各种"蹊跷"而免遭灾厄。一次是在清朝时期，当地县官本想将这古柏树砍掉。但在带着工人赶路的过程中，坐在轿子里的县太爷被一阵风吹落官帽，认为"大不祥"，此事便作罢。还有一次是在新中国成立后，当时政府在修南河大桥，计划将此树群砍伐解成木料板材。但在砍树队伍出发的那天下起了大雨，洪水堵塞河流，砍树队伍没法过河，便也"放过"了这"七贤柏"。几次蹊跷之后，古树得以存活至今。

"七贤柏"因其七株合抱独木成林，伴生七株"忠诚守卫"的奇特景观，"七—七"的神奇巧合而被赋予了各种传奇代代相传，成为当地村民们"七夕"聚会的绝佳地点。每年农历七月七，树旁边就成了庙会场所，永安、安昌等周边上万乡亲齐聚这里，上香祈福。

The Most Beautiful Cedarwood: Seven-Sage Cypress

There is a huge ancient cypress on the gentle hillside in Da'an Village of Yong'an Township, Beichuan Qiang Autonomous County in Sichuan Province. With a girth of 10.3 meters, it is second only to the ancient cypress in Huangdi Mausoleum in Shaanxi Province. It attracts many tourists.

Within a range of more than 10 meters around it, there are seven other cypress trees of 2 meters in thickness. They add radiance to each other. Because of their presence, it is also called "Seven-Sage Cypress". There are several stories about it.

According to legend, the temple abbot instructed several young monks to plant cypress trees beside the temple. However, one monk was lazy and playful. As, heavy rain was approaching, he hurriedly planted his seven cypress seedlings together. Unexpectedly, his laziness gave birth to the magical "Seven-Sage Cypress".

It survived several occurrences of strange events. One day, a county magistrate of the Qing Dynasty planned to cut it down. But his hat was blown off by a gust of wind on the way to the tree. He gave up the idea, thinking that it was very ominous. Another strange event occurred after the liberation. The local government also planned to cut down the tree to aid construction of Nanhe Bridge. However, the tree-cutting team was stopped by a sudden heavy rain, so it survived again.

The tree, with its surround by seven other small cypress trees, it is endowed with special meaning. Local villagers link it to Chinese Valentine's Day. On July 7 of the lunar calendar every year, tens of thousands of villagers from Yongan, Anchang and other surrounding areas gather around the tree, praying for blessings.

中文名：柏木
拉丁名：*Cupressus* Linn.
所在地：四川省北川羌族自治县永安镇大安村
树龄：1300 年
胸（地）围：1030 厘米
树高：3700 厘米
冠幅（平均）：1700 厘米

Chinese Name: 柏木
Latin Name: *Cupressus* Linn.
Location: Da'an Village, Yong'an Town, Beichuan Qiang Autonomous County, Sichuan Province
Tree Age: 1,300 years
Chest (Floor) Circumference: 1,030 cm
Tree Height: 3,700 cm
Crown Width (Average): 1,700 cm

最美雅安红豆树："红豆仙树"

此树位于四川省雅安市雨城区碧峰峡镇后盐村，毗邻蒙顶山和碧峰峡景区，与黄龙老鹰岩原始森林带和千佛岩作伴，形成红豆相思谷景区。

从谷歌地图俯视，红豆树处于人脸图形的耳部。红豆树有众多树根裸露石外，似瀑布飞流直下扎入大山深处。树根在生长过程中，受地形限制，经多年盘根错节生长，形似龙头。这棵神奇的红豆树，千百年来巍然屹立，直冲云霄，树干高出周围树木 10 多米，在众树中鹤立鸡群，突显奇特。

红豆树开红白相间的蝶形花，花凋谢后结出肥厚的绿豆角，秋冬季节，豆角成熟，十里八乡游人成群结队奔赴晏家山，在树下等候豆角落地。豆角落下，人们争先恐后捡拾，其乐融融。掰开豆角，里面镶嵌着象征美好爱情、成双成对的红豆果。据当地老年人讲，此树开花结果没有规律可循，有的年会开花，有的年却不开花；有时树的一半开花结果，一半不开花不结果。因此，获得红豆果的人更倍感珍稀，尤以自己亲手拾得倍觉珍贵。

红豆树主干分枝下中空。大树中心空处自下而上长了两根碗粗似藤非藤，似木非木，至今未能判断出是藤类还是树木类的植物。该植物在其基部相互缠绕，盘根错节，形成酷似小经童的状态后继续向上生长。生长数米后，植物从分枝洞口伸出，在此处树干特殊环境的塑造下，经多年后，形成一个酷似撑伞观音的造形后而终结。

红豆树因生长地理环境独特，果实形状与众不同，以及树体被大自然的鬼斧神工雕刻出各种神奇、独特形状，被当地村民称为"红豆仙树""红豆相思树"。大树底下，时有少男少女双双在此顶礼膜拜，祈求爱情也像红豆永不褪色，地久天长。

The Most Beautiful Red Bean Tree: "Red Bean Fair Tree"

The tree is located in Houyan Village of Bifengxia Township in Yucheng District, Ya'an City, Sichuan Province. It is adjacent to the Mengding Mountain Scenic Area and Bifengxia Scenic Area, which, together with Laoyingyan Primeval Forest Belt and Thousand-Buddha Crags, constitute the Red Bean Acacia Valley Scenic Area.

On the Google Maps, the tree is in the aural region of a human face. Many roots of it are exposed, looking like a waterfall flying straight down the steep mountainside. Such roots are intertwined, resembling dragon heads. The towering tree has existed for thousands of years. It is more than 10 meters higher than surrounding trees. It stands out, looking both majestic and unusual.

Its flowers are butterfly-shaped, red and white. After the flowers wither, plump green beans grow out in autumn or winter. When the beans mature, tourists from all directions rush to the Yanjia Mountain to happily pick the fallen beans. All desire to get the beans symbolizing love. According to local elders, it blooms and yields irregularly. Few beans can be harvested each year. Thus, people cherish them very much.

There is a hollow part in its trunk. Two plants like vines or wood grow out of the hollow position. They are intertwined at the roots, growing upward. They are several meters long and protrude from the branches after years of growth, finally forming a pattern of Avalokitesvara holding an umbrella.

Growing in the unique environment, it is distinctive in the shape of its fruit and body. It is a sacred tree and also "acacia rachii" in the minds of local villagers. Boys and girls are often seen praying for eternal love under the tree.

中文名：雅安红豆树
拉丁名：*Ormosia hosiei Hemsl.* etWils.
所在地：四川省雅安市雨城区碧峰峡镇后盐村
树龄：2000 年
胸（地）围：785 厘米
树高：3950 厘米
冠幅（平均）：1900 厘米

Chinese Name: 雅安红豆树
Latin Name: *Ormosia hosiei Hemsl.* etWils.
Location: Houyan Village, Bifengxia Town, Yucheng District, Ya'an City, Sichuan Province
Tree Age: 2,000 years
Chest (Floor) Circumference: 785 cm
Tree Height: 3,950 cm
Crown Width (Average): 1,900 cm

最美高山榕："华夏榕树王"

此树位于云南省盈江县铜壁关自然保护区内，有300多枝气根，入土长成新树干气根的就有168根，每年仍有10多条气根在增加，是我国目前发现的最大的榕树，因此被誉为"华夏榕树王"。

榕树王主干上布满了块状根系，由弯曲的树枝、陡直的气根构成的树洞，似殿堂；巨大的树冠，浓荫四布，遮天蔽日，像一把巨伞，伸向苍穹，好像要把蓝天白云都吞没；粗大的树干上抛撒出一束束气根，如一条条巨蟒，把头深深的扎进泥土之中，像是要把地上的水分全部吸干。整棵树，枝连枝，根连根，构成一个整体，洋洋洒洒一大片。

据说很久以前，此山林称老象坪子，坐落在叫龙盆寨的景颇族山寨。一天傍晚，山寨一景颇老人串亲回寨，拄着一根木杖，走到寨边歇脚。此时天边晚霞如血，整个山林映在霞光中。老人望着木杖，心有所思，将木杖插入土中，对杖跪拜，期盼生根发叶。深夜子时，天空电闪雷鸣，狂风暴雨。清晨朝霞染红了山林，一片春光灿烂。老人来到木杖前，被眼前的景象惊得目瞪口呆：只见昨天插入的木杖已变成粗如水桶，两丈余高，枝繁叶茂形如飞龙的榕树。老人急呼山寨村民聚于榕树下，摆放牛、羊、猪、鸡，以景颇最高礼仪祭祀。更有村民传言，曾有人不听劝告，将此树砍去一枝，回家后身体日渐消瘦，茶饭不思，不几日便气绝身亡。从此这株榕树成为当地村民的神树，无人敢攀，无人敢砍。

众多的传说更是让这棵老榕树披上了一层神秘的色彩。站在树下，人犹如依附它的一棵小草被古老的树茎包围着，仿佛能感受到老榕树默默地散发出"森林之王"的威严气息。

中文名：高山榕

拉丁名：*Ficus altissima* Bl.

所在地：云南省盈江县铜壁关自然保护区

树龄：300 年

胸（地）围：无法测量

树高：5000 厘米

冠幅（平均）：覆盖面积 9.2 亩

Chinese Name: 高山榕

Latin Name: *Ficus altissima* Bl.

Location: Tongbiguan Nature Reserve, Yingjiang County, Yunnan Province

Tree Age: 300 years

Chest (Floor) Circumference: could not be measured

Tree Height: 5,000 cm

Crown Width (Average): 0.7 hectare

The Most Beautiful Banyan: "King of Chinese Banyan"

This tree stands on Daonong Mountain near Daonong Village of Nabang Township in the Tongbiguan Nature Reserve, YingJiang County Yunnan Province. It has more than 300 aerial roots, of which 168 ones turn into new trunks. There is an increase of more than 10 aerial roots every year. It is the largest banyan tree discovered in China so far and so is hailed as the "King of Chinese Banyan".

Its trunk is covered with massive roots. The tree holes made up of curved branches and steep aerial roots ressemble palaces. Its canopy is like a huge umbrella extending toward the sky. Many aerial roots like giant pythons are intertwined around thick trunks, burying their heads deeply in the soil. The twisted roots and gnarled branches constitute a whole.

It is said that, long time ago, there was a stockaded village called Longpen inhabited by the Jingpo ethnic group. One evening, an elderly villager came to the village entrance, leaning on a wooden stick. The mountain forest was bathed in sunset glow like blood. He inserted the stick into the soil, and then bent knees, hoping it could take root and put forth leaves. At midnight, stormy weather hit the village. In the rosy dawn, he found that the wooden stick had grown into a 6.67 meters high dragon-like banyan tree with luxuriant branches and leaves. He summoned other villagers to gather under the tree and sacrifice. According to legend, one attempting to cut the tree will die of illness very soon. From then on, it has been a sacred tree in the eyes of local villagers. No one dares to climb or cut it down.

Numerous legends add more mystery to it. Standing under it, you are just like a small piece of grass embraced by ancient tree stems. It looks so majestic as the "king of forest".

The Most Beautiful Ancient Trees in China

By *Land Greening* Magazine

First English Edition 2021

By China Pictorial Press Co., Ltd.

CHINA INTERNATIONAL PUBLISHING GROUP

Copyright © China Pictorial Press Co., Ltd.

All rights reserved.

No part of this publication may be reproduced, stored in a retrieval system, or transmitted in any form or by any means, electronic, mechanical, photocopying, recording, or otherwise, without the prior written permission of China Pictorial Press Co., Ltd., except for the inclusion of brief quotations in an acknowledged review.

Address: 33 Chegongzhuang Xilu, Haidian District, Beijing, 100048, China

ISBN 978-7-5146-1990-4